Contents

D1307706

So far as the laws of
mathematics refer to reality,
they are not certain.
And so far as they are certain,
they do not refer to reality.

– Albert Einstein

From almost naught to almost all I flee,
and almost has almost confounded me
zero my limit, and infinity!

On the Calculus,
– W. Cummings

About the Author

James D. Broesch is a staff engineer at General Atomics. He is responsible for the design and development of several advanced control systems used on the DIII-D Tokamak Fusion Research Program. He has over 10 years of experience in designing and developing communications and control systems for applications ranging from submarines to satellites. Mr. Broesch teaches classes in signal processing and hardware design via the Extension office of the University of California, San Diego. He is the author of *Practical Programmable Circuits* (Academic Press) as well as magazine articles and numerous technical papers.

Dedication

This book is dedicated to the Dubel family
for their many years of friendship.

Acknowledgments

In any project as large as this book there are a number of people who, directly or indirectly, make major contributions. I would like to thank Dr. Mike Walker of General Atomics for his assistance and guidance, and for reviewing the final manuscript. I would also like to thank Dr. Mark Kent for his help with the initial concept and early development. Riley Woodson's input on the initial draft of the manuscript, and on the user interface of the DSP Calculator software, was insightful and of great value.

I would also like to thank Tony Bowers of Quest-Rep and Bruce Newgard of Xilinx. Their technical support of my design work, and their willingness to share their knowledge and skills in the classes I teach, have contributed to my knowledge and understanding of digital signal processing. Finally, I would like to thank the students in the UCSD Extension Program and the other readers who took the time to point out typos, suggest improvements, and otherwise helped make this a better work.

About the Accompanying Software

What is the *DSP Calculator*?

The *DSP Calculator* suite of software routines is designed to illustrate many of the basic concepts involved in working with DSP. The tools included with the *DSP Calculator* enable you to create waveforms, design filters, filter the waveforms, and display the results. Also included are routines for generating complex waveforms based on the complex exponential, routines that perform the discrete Fourier transform (DFT), and routines for computing the more computationally efficient fast Fourier transform (FFT). Several experiments that make use of the *DSP Calculator* are included within the text of this book. These are indicated with the following graphic symbol:

Interactive Exercise

The programs can be, and have been, used to develop practical commercial applications. They are not, however, intended for developing large-scale or critical DSP implementations. Their purpose is primarily educational. Use them to experiment with the DSP concepts introduced in this book; you'll quickly develop an intuitive sense for the math behind the concepts!

The *DSP Calculator* software runs under Microsoft Windows 3.1 or later. It is designed to run on a 386 system with at least 4 Mbytes of RAM. A math co-processor is not required, but is recommended. The programs have been tested under Windows 95 with both 486 and Pentium processors.

About This Manual

This manual describes the installation, use, and data formats for each of the programs that make up the *DSP Calculator*. Examples are given for each of the programs. This manual assumes that the user is familiar with the concepts developed in the text. If something does not make sense to you, please refer to the appropriate section in the text.

Installation and General Information

Each program in the *DSP Calculator* suite is designed to load or save data in a standard file format. This makes it easy to use the programs in combination with each other. The file format also makes interchanging data with other programs relatively straightforward.

All data is stored as ASCII text in a "comma delimited format." The comma separates the real part of the number from the imaginary part. All data is stored as complex floating-point numbers. If the data has only real values, then the imaginary part will be zero. If the data has only imaginary values, then the real part will be set to zero. All numbers are floating point values, though a number may be expressed as an integer if it does not have any values in the decimal place.

This is best illustrated with an example. The numbers 1, 2.3, 3.0, 4.3+j1, j5, would be stored in a file as follows:

$$1, 0$$
$$2.3, 0$$
$$3, 0$$
$$4.3, 1$$
$$0, 5.$$

Notice that each number is on a line by itself. This format allows data to be manipulated with a standard text editor, to be read or written easily by C, BASIC, or FORTRAN programs, or to be easily interchanged with spreadsheets or math programs.

When first invoked, all programs come up with a reasonable set of default values for the program's parameters. Typical parameters include amplitude, frequency, and the number of samples.

Two assumptions are made about all data. First, it is assumed that all data is uniformly sampled, or, in other words, that the time interval between all samples is the same. The other assumption is that all angle data is in radians. Thus, all frequency graphs are shown having values between $-\pi$ and π. The actual frequency is related to the sample rate by the equation:

$$\pi = \frac{f_s}{2}$$

For example, if the number of samples (f_s) is 100 samples/second, then a frequency of π is equivalent to 50 Hz. For this example, a frequency of $\pi/2$ would be equal to 25 Hz, and so forth. Notice that no time units are given in *DSP Calculator*. Only the number of samples are used.

As noted above, all parameters are set at startup with reasonable values. It is possible, however, to generate output that cannot be properly displayed. One of two things will happen in this case:

a. The display will simply look strange, or

b. A message box will be generated that warns that the data cannot be displayed correctly.

In either case, the data in the program's buffers will be correct. Even if the data is not displayed correctly, the data in the buffers can still be saved to a file.

The practical limitations of the *DSP Calculator* should be kept in mind when using it with other programs. In general, the maximum number of samples that can be placed in a file is 10,240. Some modules have other restrictions. The FFT program, for example, is restricted to a maximum 1024 samples, and the number of samples must be even.

It should be kept in mind when using the *DSP Calculator* that its primary purpose is educational. Thus, it will accept input values that commercial design programs might block. For example, both negative amplitudes and negative frequencies are accepted, and the data generated accordingly. Engineers do not normally think about amplitudes or frequencies as being negative, but these values are not merely mathematical abstractions. A negative amplitude simply means that the signal is inverted from an equivalent positive amplitude; a negative frequency relates to the phase of the signal. See the section in the text on complex numbers for a thorough discussion of negative frequencies.

Fungen

Purpose: This is the general-purpose function generator. It will produce sine waves, square waves, and triangle waveforms.

Inputs: There are five parameters that can be set:

Frequency
Amplitude
Offset
Phase
Number of Samples

There are four buttons:

Sin

Square

Triangle

Clear

Most of these are self-explanatory. The number of samples must be less than 10,241. The Clear button clears the screen and erases the internal buffer.

Outputs: The waveform displayed is kept in an internal buffer. This buffer can be saved to a file by using the FILE / SAVE option on the menu bar.

Operation: Using Fungen is straightforward. Simply enter the desired parameters, then press the appropriate button. The waveform will be shown on the screen, and the data will be saved in the internal buffer. Each time one of the function buttons is pushed, the internal buffer is erased and a new waveform is generated and stored. This makes it easy to adjust parameters: simply change the desired parameter and hit the function button again.

Example: From the DSPCALC folder, double-click on the Fungen icon. The function generator will appear. Using the mouse, click on the Sin button. Two cycles of a sine waveform will be shown. Save this waveform to a file by clicking on the FILE menu. Then click on the SAVE button. Enter a file name such as EXAMPL1.SIG and then click on the OK button. Open the file using Window's Notepad application. Assuming that you have used the standard installation path, the path name will be DSPCALC/FUNGEN/EXAMPL1.SIG. You will see the numeric values for the waveform.

Fourier

Purpose: Fourier is used for two purposes: It demonstrates the concept of building up a waveform from simple sine waves. Secondly, it is used to create test waveforms for the filter functions.

Inputs: There are three parameters that can be set:

Frequency

Amplitude

Number of Samples

There are three buttons:

Sin

Cos

Clear Screen

Most of these are self-explanatory. The number of samples must be less than 10,241. The Clear Screen button clears the screen and erases the internal buffer.

Outputs: The waveform displayed is kept in an internal buffer. This buffer can be saved to a file by using the FILE / SAVE option on the menu bar.

Operation: To use Fourier enter the desired parameters, and then press the appropriate button. The waveform will be shown on the screen, and the data will be saved in the internal buffer. Unlike the function generator, each time one of the function buttons is pushed the internal buffer is not erased. The new waveform component is added to the buffer and the waveform is displayed.

Example: From the DSPCALC folder, double-click on the Fourier icon. When the Fourier window appears, click on the Sin button. You will see a waveform appear on the screen. Now change the value of the frequency to 6. Next change the value of the amplitude to 0.3333. Click on the Sin button again. Notice that the new waveform is a composite. Finally, change the value of the frequency to 10 and the value of the amplitude to 0.2 and click on Sin again. These values correspond to the first three terms in the Fourier series for a square wave, so the resulting waveform should begin to look like a square wave with rounded corners. Save the file under the name EXAMPL2.SIG.

DFT

Purpose: DFT is used to convert a signal in the time domain to a signal in the frequency domain. It is similar to FFT. The Discrete Fourier Transform, however, is more flexible and should be used whenever the transform of a complex series is required.

Inputs: There are two parameters that can be set:

Amplitude
Number of Samples

There are two buttons:

Transform
Refresh

The Amplitude dialog adjusts the amplitude of the signal display. It does not affect the signal itself—only the display is affected. The number of samples is limited to 1024. Transform performs the DFT on the signal. Refresh is used to redraw the screen, if necessary. This can be handy if other windows have erased part of the screen display.

Outputs: The transformed waveform displayed is kept in an internal buffer. This buffer can be saved to a file by using the FILE / SAVE option on the menu bar.

Operation: Use the FILE / LOAD menu to load the signal. The correct number of samples and the amplitude should be set before the file is loaded. The waveform will be displayed. Click on the waveform, and the display will change to show the spectrum of the signal. The transformed signal can be saved using the FILE / SAVE option on the menu bar.

Example: This example assumes that the file EXAMPL2.SIG exists. The EXAMPL2.SIG file was created in the example on the use of the Fourier program. Use the FILE / LOAD menu to load the file /DSPCALC/FOURIER/EXAMPL2.SIG.

FFT

The FFT program is similar to the DFT program. It uses a computationally efficient FFT algorithm to obtain the transform, however. The FFT routine is considerably faster, but it is restricted to handling sample counts that are a power of 2. The sample count must be 2, 4, 8, 16, 32, 64, 128, 256, 512, or 1024.

Example: Transform the signal file EXAMPLE2.SIG, as described in the discussion of the DFT program. You should get the same results. However, the program will execute in much less time than DFT.

Cmplxgen

Purpose: The purpose of this program is to generate complex waveforms based on the equation e^s where $s = \alpha + j2\pi f$.

Inputs: There are four parameters that can be set:

Frequency
Amplitude
Alpha
Number of Samples

There are two action buttons:

Generate

Clear Screen

The Generate button computes and displays the waveform. The Clear Screen button clears the display and erases the internal buffer. The number of samples is limited to 10,240.

Outputs: The waveform displayed is kept in an internal buffer. This buffer can be saved to a file by using the FILE / SAVE option on the menu bar.

Operation: Enter the desired parameters and click on the Generate button.

Example: Double-click on the Cmplxgen icon. Then click on the Generate button. The unit circle will be plotted on the left-hand side of the screen. The corresponding real and imaginary plots will be generated on the right-hand side of the screen.

REDISP

Purpose: REDISP is a general-purpose display program. It will display the real portion of waveforms stored in the DSPCALC format signal files.

Inputs: There are two parameters that can be set:

Amplitude

Number of Samples

There are no action buttons. There is, however, a *frame* slider located at the bottom of the window. Please see the discussion under "Operation."

Outputs: There are no outputs other than the display.

Operation: Enter the desired amplitude and the number of samples for the signal that will be displayed. Use the FILE / LOAD menu to select the file to display. The display is divided into ten frames. Each frame can display up to 1024 samples.

Example: Use Cmplxgen to produce a signal with the following parameters:

$$Frequency = 20$$
$$Amplitude = 1$$
$$Alpha = -2$$
$$Number\ of\ samples = 10,240$$

Save this file as EXAMPL3.SIG. Then invoke REDISP by double-clicking on its icon. Set the number of samples to 1024. Then load the EXAMPL3.SIG file using the FILE / LOAD menu. Notice that the first few cycles of the signal are shown. Display the rest of the waveform by using the slider at the bottom of the screen.

IMDISP

IMDISP is similar to REDISP. The only difference is that IMDISP displays the imaginary portion of the waveform.

CONVOLVE

Purpose: This program performs the convolution of two data sequences. Each data sequence is stored in its own file.

Inputs: There are two parameters that can be set:

$$Amplitude$$
$$Number\ of\ Samples$$

There are two action buttons and a *frame* slider located at the bottom of the window. Please see the discussion under "Operation."

Outputs: The result of convolving the two sequences is shown on the screen and saved in the internal buffer. This data can be saved to a file using the FILE menu.

Operation: *Convolve* is normally used to perform some type of filtering operation. Enter the desired amplitude and the number of samples for the signal that will be displayed. Use the FILE / LOAD COEFFICIENTS menu to select the coefficients to use. The coefficients waveform will be displayed. Next, use the FILE / LOAD SIGNAL menu to load in the signal. The *Convolve* command button will cause the signal to be convolved through the coefficients.

The result of the convolution will be displayed. The display is divided into ten frames. Each frame can display up to 1024 samples.

Example: See the text on FIR filtering for a detailed example of using *Convolve*.

FLTRDSGN

Purpose: This program is used to design filters. More specifically, it is used to produce the coefficients for low-pass, bandpass, or high-pass filters.

Inputs: The inputs to this program depend upon the type of filter being designed. There are no action buttons on the main screen.

Outputs: The coefficients for the filter are saved in the internal buffer. These can be saved to a file by using the FILE / SAVE AS menu.

Operation: The type of filter to be designed is selected using the FILTERS menu. The three selections are FILTERS / LOW PASS, FILTERS / BAND PASS, FILTERS / HIGH PASS.

Example: Design a 33-tap bandpass filter that will pass signals from $\pi/4$ to $3\pi/4$.

First, select the FILTERS / BAND PASS menu. A window will appear with boxes for the lower cutoff frequency, the upper cutoff frequency, and the number of taps for the filter. In the lower cutoff frequency box enter 0.785 ($\pi/4$). In the upper cutoff frequency box enter 2.36 ($3\pi/4$). Then enter 33 into the Number of Taps box.

Next, press the OK button. The frequency response curve for the filter will be displayed. You can experiment with the shape of the curve by changing the number of taps.

Preface

Digital signal processing (DSP) is one of the fastest-growing fields in modern electronics. Only a few years ago DSP techniques were considered advanced and esoteric subjects, their use limited to research labs or advanced applications such as radar identification. Today, the technology has found its way into virtually every segment of electronics. Talking toys, computer graphics, and CD players are just a few of the common examples.

The rapid acceptance and commercialization of this technology has presented the modern design engineer with a serious challenge: either gain a working knowledge of the new techniques or risk obsolescence. Unfortunately, anyone attempting to gain this knowledge has had to face some serious obstacles. Traditionally, engineers have had two options for acquiring new skills: go back to school, or turn to vendor's technical documentation. In the case of DSP, neither of these approaches is a particularly good one.

Undergraduate programs—and even most graduate programs— devoted to DSP are really only thinly disguised courses in the mathematical discipline known as *complex analysis*. The purpose of most college programs is *not* to teach a working knowledge of DSP; the purpose of these programs is to prepare students for graduate research on DSP topics. Many subjects such as the Laplace transformation, even and odd functions, and so forth are covered in depth, while much of the information needed to really comprehend the "whys and wherefores" of DSP techniques are left unmentioned.

Manufacturer documentation is often of little more use to the uninitiated. Applications notes and design guides usually are either reprints of textbook discussions, or they focus almost exclusively on particular features of the vendor's instruction set or architecture.

The purpose of this book is to bridge the gap between the theory of digital signal processing and the practical knowledge necessary to understand a working DSP system. The mathematics is not ignored; you will see many sophisticated mathematical relationships in thumbing through the pages of this work. What is left out, however, are the formal proofs, the esoteric discussions, and the tedious mathematical exercises. In their place are thorough background discussions explaining how and why the math is important, examples that can be run on any general-purpose computer, and tips that can help you gain a comfortable understanding of the DSP processes.

This book is specifically written for the working engineer, but many others can benefit from the material contained here. Program managers that find they need to understand DSP concepts will appreciate the straightforward presentation. Students who are about to embark on formal DSP programs will find this information useful as a gentle introduction to an intimidating subject. Those students who have had formal DSP training, but feel a lack of clear understanding, will find that this book provides a convenient place to clear up many fuzzy concepts.

While the material is written for engineers, the mathematics is kept as simple as possible. A first-year course in trigonometry combined with a first-year course in calculus will provide more than adequate preparation. Even those engineers who have been away from the books for a while should have no difficulty in following the mathematics. Special care is taken throughout to introduce all mathematical discussions and, since formal proofs are not presented, few esoteric relationships need to be mastered.

CHAPTER 1

Digital Signal Processing

The Need for DSP

What is digital signal processing (DSP) anyway, and why should we use it? Before discussing either the hardware, the software, or the underlying mathematics, it's a good idea to answer these basic questions.

The term DSP generally refers to the use of digital computers to process signals. Normally, these signals can be handled by analog processes but, for a variety of reasons, we may prefer to handle them digitally.

To understand the relative merits of analog and digital processing, it is convenient to compare the two techniques in a common application. Figure 1-1 shows two approaches to recording sounds such as music or speech. Figure 1-1a is the analog approach. It works like this:

- Sound waves impact the microphone, where they are converted to electrical impulses.

- These electrical signals are amplified, then converted to magnetic fields by the recording head.

- As the magnetic tape moves under the head, the intensity of the magnetic fields is stored on the tape.

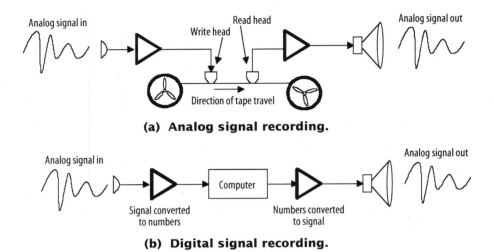

(a) Analog signal recording.

(b) Digital signal recording.

Figure 1-1: Analog and digital systems.

The playback process is just the inverse of the recording process:

- As the magnetic tape moves under the playback head, the magnetic field on the tape is converted to an electrical signal.

- The signal is then amplified and sent to the speaker. The speaker converts the amplified signal back to sound waves.

The advantage of the analog process is twofold: first, it is conceptually quite simple. Second, by definition, an analog signal can take on virtually an infinite number of values within the signal's dynamic range. Unfortunately, this analog process is inherently unstable. The amplifiers are subject to gain variation over temperature, humidity, and time. The magnetic tape stretches and shrinks, thus distorting the recorded signal. The magnetic fields themselves will, over time, lose some of their strength. Variations in the speed of the motor driving the tape cause additional distortion. All of these factors combine to ensure that the output signal will be considerably lower in quality than the input signal. Each time the signal is passed on to another analog process, these adverse effects are multiplied. It is rare for an analog system to be able to make more than two or three generations of copies.

Now let's look at the digital process as shown in Figure 1-1b:

- As in the analog case, the sound waves impact the microphone and are converted to electrical signals. These electrical signals are then amplified to a usable level.

- The electrical signals are measured or, in other words, they are converted to numbers.

- These numbers can now be stored or manipulated by a computer just as any other numbers are.

- To play back the signal, the numbers are simply converted back to electrical signals. As in the analog case, these signals are then used to drive a speaker.

There are two distinct disadvantages to the digital process: first, it is far more complicated than the analog process; second, computers can only handle numbers of finite resolution. Thus, the (potentially) "infinite resolution" of the analog signal is lost.

Advantages of DSP

Obviously, there must be some compensating benefits of the digital process, and indeed there are. First, once converted to numbers, the signal is unconditionally stable. Using techniques such as error detection and correction, it is possible to store, transmit, and reproduce numbers with no corruption. The twentieth generation of recording is therefore just as accurate as the first generation.

This fact has some interesting implications. Future generations will never really know what the Beatles sounded like, for example. The commercial analog technology of the 1960s was simply not able to accurately record and reproduce the signals. Several generations of analog signals were needed to reproduce the sound: First, a master tape would be recorded, and then mixed and edited; from this, a metal master record would be produced, from which would come a plastic impression. Each step of the process was a new generation of recording, and each generation acted on the signal like a filter, reducing the frequency content and skewing the phase. As with the paintings in the Sistine Chapel, the true colors and brilliance of the original art is lost to history.

Things are different for today's musicians. A thousand years from now historians will be able to accurately play back the digitally mastered CDs of today. The discs themselves may well deteriorate, but before they do, the digital numbers on them can be copied with perfect accuracy. Signals stored digitally are really just large arrays of numbers. As such, they are immune to the physical limitations of analog signals.

There are other significant advantages to processing signals digitally. Geophysicists were one of the first groups to apply the techniques of signal processing. The seismic signals of interest to them are often of very low frequency, from 0.01 Hz to 10 Hz. It is difficult to build analog filters that work at these low frequencies. Component values must be so large that physically implementing the filter may well be impossible. Once the signals have been converted to digital numbers, however, it is a straightforward process to program a computer to perform the filtering.

Other advantages to digital signals abound. For example, DSP can allow large bandwidth signals to be sent over narrow bandwidth channels. A 20-kHz signal can be digitized and then sent over a 5-kHz channel. The signal may take four times as long to get through the narrower bandwidth channel, but when it comes out the other side it can be reconstructed to its full 20-kHz bandwidth.

In the same way, communications security can be greatly improved through DSP. Since the signal is sent as numbers, it can be easily encrypted. When received, the numbers are decrypted and then reproduced as the original signal. Modern "secure telephone" DSP systems allow this processing to be done with no detectable effect on the conversation.

Chapter Summary

Digitally processing a signal allows us to do things with signals that would be difficult, or impossible, with analog approaches. With modern components and techniques, these advantages can often be realized economically and efficiently.

CHAPTER 2

The General Model of a DSP System

Introduction

The general model for a DSP system is shown in Figure 2-1. From a high-level point of view, a DSP system performs the following operations:

- Accepts an analog signal as an input.

- Converts this analog signal to numbers.

- Performs computations using the numbers.

- Converts the results of the computations back into an analog signal.

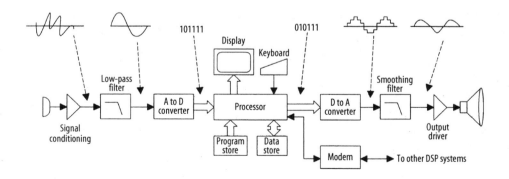

Figure 2-1: The general model for a DSP system.

Optionally, different types of information can be derived from the numbers used in this process. This information may be analyzed, stored, displayed, transmitted, or otherwise manipulated.

This model can be rearranged in several ways. For example, a CD player will not have the analog input section. A laboratory instrument may not have the analog output. The truly amazing thing about DSP systems, however, is that the model will fit *any* DSP application. The system could be a sonar or radar system, voicemail system, video camera, or a host of other applications. The specifications of the individual key elements may change, but their function will remain the same.

In order to understand the overall DSP system, let's begin with a qualitative discussion of the key elements.

Input

All signal processing begins with an *input transducer*. The input transducer takes the input signal and converts it to an electrical signal. In signal-processing applications, the transducer can take many forms. A common example of an input transducer is a microphone. Other examples are geophones for seismic work, radar antennas, and infrared sensors. Generally, the output of the transducer is quite small: a few microvolts to several millivolts.

Signal-conditioning Circuit

The purpose of the signal-conditioning circuit is to take the few millivolts of output from the input transducer and convert it to levels usable by the following stages. Generally, this means amplifying the signal to somewhere between 3 and 12V. The signal-conditioning circuit also limits the input signal to prevent damage to following stages. In some circuits, the conditioning circuit provides isolation between the transducer and the rest of the system circuitry.

Typically, signal-conditioning circuits are based on operational amplifiers or instrumentation amplifiers.

Anti-aliasing Filter

The anti-aliasing filter is a low-pass filter. The job of the anti-aliasing filter is a little difficult to describe without more theoretical background than we have developed up to this point (see Chapter 6 for more details). However, from a conceptual point of view, the anti-aliasing filter can be thought of as a mechanism to limit how fast the input signal can change. This is a critical function; the anti-aliasing filter ensures that the rest of the system will be able to track the signal. If the signal changes too rapidly, the rest of the system could miss critical parts of the signal.

Analog-to-Digital Converter

As the name implies, the purpose of the analog-to-digital converter (ADC) is to convert the signal from its analog form to a digital data representation. Due to the physics of converter circuitry, most ADCs require inputs of at least several volts for their full range input. Two of the most important characteristics of an ADC are the *conversion rate* and the *resolution*. The conversion rate defines how fast the ADC can convert an analog value to a digital value. The resolution defines how close the digital number is to the actual analog value.

The output of the ADC is a binary number that can be manipulated mathematically.

Processor

Theoretically, there is nothing special about the processor. It simply performs the calculations required for processing the signal. For example, if our DSP system is a simple amplifier, then the input value is literally multiplied by the gain (amplification) constant.

In the early days of signal processing, the processor was often a general-purpose mainframe computer. As the field of DSP progressed, special high-speed processors were designed to handle the "number crunching."

Today, a wide variety of specialized processors are dedicated to DSP. These processors are designed to achieve very high data throughputs, using a combination of high-speed hardware, specialized architectures, and dedicated instruction sets. All of these functions are designed to efficiently implement DSP algorithms.

Program Store, Data Store

The *program store* stores the instructions used in implementing the required DSP algorithms. In a general-purpose computer (von Neumann architecture), data and instructions are stored together. In most DSP systems, the program is stored separately from the data, since this allows faster execution of the instructions. Data can be moved on its own bus at the same time that instructions are being fetched. This architecture was developed from basic research performed at Harvard University, and therefore is generally called a *Harvard architecture*. Often the data bus and the instruction bus have different widths.

Data Transmission

DSP data is commonly transmitted to other DSP systems. Sometimes the data is stored in bulk form on magnetic tape, optical discs (CDs), or other media. This ability to store and transmit the data in digital form is one of the key benefits of DSP operations. An analog signal, no matter how it is stored, will immediately begin to degrade. A digital signal, however, is much more robust since it is composed of ones and zeroes. Furthermore, the digital signal can be protected with error detection and correction codes.

Display and User Input

Not all DSP systems have displays or user input. However, it is often handy to have some visual representation of the signal. If the purpose of the system is to manipulate the signal, then obviously the user needs a way to input commands to the system. This can be accomplished with a specialized keypad, a few discrete switches, or a full keyboard.

Digital-to-Analog Converter

In many DSP systems, the signal must be converted back to analog form after it has been processed. This is the function of the digital-to-analog converter (DAC). Conceptually, DACs are quite straightforward: a binary number put on the input causes a corresponding voltage on the output. One of the key specifications of the DAC is how fast the output voltage settles to the commanded value. The slew rate of the DAC should be matched to the acquisition rate of the ADC.

Output Smoothing Filter

As the name implies, the purpose of the smoothing filter is to take the edges off the waveform coming from the DAC. This is necessary since the waveform will have a "stair-step" shape, resulting from the sequence of discrete inputs applied to the DAC.

Generally, the smoothing filter is a simple low-pass system. Often, a basic RC circuit does the job.

Output Amplifier

The output amplifier is generally a straightforward amplifier with two main purposes. First, it matches the high impedance of the DAC to the low impedance of the transducer. Second, it boosts the power to the level required.

Output Transducer

Like the input transducer, the output transducer can assume a variety of forms. Common examples are speakers, antennas, and motors.

Chapter Summary

The overall idea behind digital signal processing is to:

- Acquire the signal.

- Convert it to a sequence of digital numbers.

- Process the numbers as required.

- Transmit or save the data as may be required.

- Convert the processed sequence of numbers back to a signal.

This process may be considerably more complicated than the traditional analog signal processors (radios, telephones, TVs, stereos, etc.) However, given the advances in modern technology, DSP solutions can be both cheaper and far more efficient than traditional techniques.

This chapter has looked at the key blocks in a DSP system. Any DSP system will be composed of some subset of these blocks. The key to understanding, specifying, or designing a DSP system is to know how these blocks are related, and how the parameters of any one block impact the parameters of the other blocks. The rest of this book is dedicated to providing this level of understanding.

CHAPTER **3**

The Numerical Basis for DSP

Introduction

The heart of DSP is, naturally enough, numbers. More specifically, DSP deals with how numbers are processed. Most texts on DSP either assume that the reader already has a background in numerical theory, or they add an appendix or two to review complex numbers. This is unfortunate, since the key algorithms in DSP are virtually incomprehensible without a strong foundation in the basic numerical concepts.

Since the numerical foundation is so critical, we begin our discussion of the mathematics of DSP with some basic information. This material may be review, especially for those readers who are well versed in trigonometry. However, we suggest that you at least scan the material presented in this section, as the discussions that follow this section will be much clearer. Also, Appendix A reviews some of the fundamentals of engineering calculus and other mathematical tools.

In general, applied mathematics is a study of functions. Primarily, we are interested in how the function behaves directly. That is, for any given input, we want to know what the output is. Often, however, we are interested in other properties of a given function. For example, we may want to know how rapidly the function is changing, what the maximum or minimum values are, or how much area the function bounds.

Additionally, it is often handy to have a couple of different ways to express a function. For some applications, one expression may make our work simpler than another.

Polynomials, Transcendental Functions, and Series Expansions

Polynomials are the workhorse of applied mathematics. The simplest form of the polynomial is the simple linear equation:

$$y = mx + b$$

Equation 3-1

where m and b are constants. For any straight line drawn on an x-y graph, an equation in the form of Equation 3-1 can be found. The constant m defines the slope, and b defines the y-intercept point. Not all functions are straight lines, of course. If the graph of the function has some curvature, then a higher-order function is required. In general, for any function, a polynomial can be found of the form:

$$f(x) = ax^n + \ldots + bx^1 + cx^0$$

Equation 3-2

which closely approximates the given function, where a, b, and c are constants called the coefficients of $f(x)$.

This polynomial form of a function is particularly handy when it comes to differentiation or integration. Simple arithmetic is normally all that is needed to find the integral or derivative. Furthermore, computing a value of a function when it is expressed as a polynomial is quite straightforward, particularly for a computer.

If polynomials are so powerful and easy to use, why do we turn to *transcendental* functions such as the sine, cosine, natural logarithm, and so on? There are a number of reasons why transcendental functions are useful to us. One reason is that the transcendental forms are simply more compact. It is much easier to write:

$$y = \sin(x) \qquad \text{Equation 3-3}$$

than it is to write the polynomial approximation:

$$f(x) = x - \frac{1}{3!}x^3 + \frac{1}{5!}x^5 - \dots \qquad \text{Equation 3-4}$$

Another reason is that it is often much easier to explore and manipulate relationships between functions if they are expressed in their transcendental form.

For example, one look at Equation 3-3 tells us that $f(x)$ will have the distinctive shape of a sine wave. If we look at Equation 3-4, it's much harder to discern the nature of the function we are working with. It is worth noting that, for many practical applications, we do in fact use the polynomial form of the function and its transcendental form interchangeably. For example, in a spreadsheet or high-level programming language, a function call of the form:

$$y = \sin(x) \qquad \text{Equation 3-5}$$

results in y being computed by a polynomial form of the sine function.

Often, polynomial expressions called *series expansions* are used for computing numerical approximations. One of the most common of all series is the Taylor series. The general form of the Taylor series is:

$$f(x) = \sum_{n=0}^{\infty} a_n x^n \qquad \text{Equation 3-6}$$

Again, by selecting the values of a_n, it is possible to represent many functions by the Taylor series. In this book we are not particularly interested in determining the values of the coefficients for functions in general, as this topic is well covered in many books on basic calculus. The idea of series expansion is presented here because it plays a key role in an upcoming discussion: the z-transform.

A series may converge to a specific value, or it may diverge. An example of a convergent series is:

$$f(n) = \sum_{n=0}^{\infty} \frac{1}{2^n} \qquad\qquad \text{Equation 3-7}$$

As n grows larger, the term $1/2^n$ grows smaller. No matter how many terms are evaluated, the value of the series simply moves closer to a final value of 2.

A divergent series is easy to come up with:

$$f(n) = \sum_{n=0}^{\infty} 2^n \qquad\qquad \text{Equation 3-8}$$

As n approaches infinity, the value of $f(n)$ grows without bound. Thus, this series diverges.

It is worth looking at a practical example of the use of series expansions at this point. One of the most common uses of series is in situations involving growth. The term *growth* can be applied to either biological populations (herds, for example), physical laws (the rate at which a capacitor charges), or finances (compound interest).

Let's take a look at the concept of compound growth. The idea behind it is simple:

- You deposit your money in an account.

- After some set period of time (say, a month), your account is credited with interest.

- During the next period, you earn interest on both the principal *and* the interest from the last period.

- This process continues as described above.

Your money keeps growing at a faster rate, since you are earning interest on the previous interest as long as you leave the money in the account.

Mathematically, we can express this as:

$$f(x) = x + \frac{x}{c}$$
  Equation 3-9

where c is the interest rate. If we start out with a dollar, and have an interest rate of 10% per month, we get:

$$f(1) = 1 + \frac{1}{10}$$
$$= 1.10$$

for the first month. For the second month, we would be paid interest on $1.10:

$$f(1.10) = 1.10 + \frac{1.10}{10}$$
$$= 1.21$$

and so on. This type of computation is not difficult with a computer, but it can be a little tedious. It would be nice to have a simple expression that would allow us to compute what the value of our money would be at any given time. With some factoring and manipulation, we can come up with such an expression:

$$f(n) = \left(x + \frac{x}{c} \right)^n$$

Equation 3-10

where n is the number of compounding periods. Using Equation 3-10 we can directly evaluate what our dollar will be worth after two months:

$$f(2) = \left(1 + \frac{1}{10} \right)^2$$
$$= 1.1^2$$
$$= 1.21$$

For many applications, the value of c is proportional to the number of periods. For example, when a capacitor is charging, it will reach half its value in the first time period. During the next time period, it will take on half of the previous value (that is $1/4$), etc. For this type of growth, we can set $c = n$ in Equation 3-10. Assuming a starting value of 1, we get an equation of the following form:

$$f(n) = \left(1 + \frac{1}{n} \right)^n$$

Equation 3-11

Equation 3-11 is a *geometric series*. As n grows larger, $f(n)$ converges to the irrational number approximated by 2.718282. (You can easily verify this with a calculator or spreadsheet.) This number comes up so often in mathematics that is has been given its own name: *e*. Using *e* as a base in logarithm calculations greatly simplifies problems involving this type of growth. The natural logarithm (ln) is defined from this value of *e:*

$$\ln(e) = 1 \qquad \textbf{Equation 3-12}$$

It is worth noting that the function e^x can be rewritten in the form of a series expansion:

$$e^x = 1 + x + \frac{x^2}{2!} + \dots \frac{x^n}{n!} + \dots \qquad \textbf{Equation 3-13}$$

The natural logarithm and the base *e* play an important role in a wide range of mathematical and physical applications. We're primarily interested in them, however, for their role in the use of imaginary numbers. This topic will be explored later in this chapter.

Limits

Limits play a key role in many modern mathematical concepts. They are particularly important in studying integrals and derivatives. They are covered here mainly for completeness of this discussion.

The basic mathematical concept of a limit closely parallels what most people think of as a limit in the physical world. A simple example is a conventional signal amplifier. If our input signal is small enough, the output will simply be a scaled version of the input. There is, however, a limit to how large an output signal we can achieve. As the amplitude of the input signal is increased, we will approach this limit. At some point, increasing the amplitude of the input will make no difference on the output signal; we will have reached the limit.

Mathematically, we can express this as:

$$v_{out_{max}} = \lim_{x \to v_{in_{max}}} f(x) \qquad \text{Equation 3-14}$$

where $f(x)$ is the output of the amplifier, and $v_{in_{max}}$ is the maximum input voltage that does not cause the amplifier to saturate.

Limits are often evaluated under conditions that make mathematical sense, but do not make intuitive sense to most us. Consider, for example, the function $f(x) = 2 + 1/x$. We can find the value of this function as x takes on an infinite value:

$$\lim_{x \to \infty} \left(2 + \frac{1}{x} \right) = 2$$

In practice, what we are saying here is that as x becomes infinitely large, then $1/x$ becomes infinitely small. Intuitively, most people have no problem with dropping a term when it no longer has an effect on the result. It is worth noting, however, that mathematically the limit is not just dropping a noncontributing term; the value of 2 is a mathematically precise solution.

Integration

Many concepts in DSP have geometrical interpretations. One example is the geometrical interpretation of the process of integration. Figure 3-1 shows how this works. Let's assume that we want to find the area under the curve $f(x)$. We start the process by defining some handy interval—in this case, simply $b - a$. This value is usually defined as Δx. For our example, the interval Δx remains constant between any two points on the x-axis. This is not mandatory, but it does make things easier to handle.

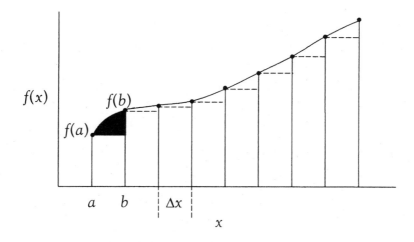

Figure 3-1: Geometric interpretation of integration.

Now, integration is effectively a matter of finding the area under the curve $f(x)$. A good approximation for the area in the region from a to b and under the curve can be found by multiplying $f(a)$ by Δx. Mathematically:

$$\int_{a}^{b} f(x)dx \approx f(a)\Delta x \qquad \text{Equation 3-15}$$

Our approximation will be off by the amount between the top of the rectangle formed by $f(a)\Delta x$ and yet still under the curve $f(x)$. This is shown as a shaded region in Figure 3-1. For the interval from a to b this error is significant. For some of the other regions this error can be seen to be insignificant. The overall area under the curve is the sum of the individual areas:

$$\int f(x)dx \approx \sum f(x)\Delta x \qquad \text{Equation 3-16}$$

It's worthwhile to look at the source of error between the integral and our approximation. If you look closely at Figure 3-1, you can see that the major factor determining the error is the size of Δx. The smaller the value of Δx, the closer the actual value of the integral and our approximation will be. In fact, if the value of Δx is made vanishingly small, then our approximation would be exact. We can do this mathematically by taking the limit of the right-hand side of Equation 3-16 as Δx approaches 0:

$$\int f(x)dx = \lim_{\Delta x \to 0} \sum f(x)\Delta x \qquad \text{Equation 3-17}$$

Notice that Equation 3-17 is in fact the definition of the integral, not an approximation.

There are a number of ways to find the integral of a function. Numerically, a value can be computed using Equation 3-16 or some more sophisticated approximation technique. For symbolic analysis, the integral can be found by using special relationships or, as is more often the case, by tables. For most DSP work, only a few simple integral relationships need to be mastered. Some of the most common integrals are shown in Table A.3 of Appendix A.

Oscillatory Motion

Virtually all key mathematical concepts in DSP can be directly derived from the study of oscillatory motion. In physics, there are a number of examples of oscillatory motion: weights on springs, pendulums, LC circuits, etc. In general, however, the simplest form of oscillatory motion is the wheel. Think of a point on the rim of a wheel. Describe how the point on the wheel moves mathematically and the foundations of DSP are in place. This statement may seem somewhat dramatic, but it is truly amazing how often this simple fact is overlooked.

The natural place to begin describing circular motion is with Cartesian coordinates. Figure 3-2 shows the basic setup. The origin of the coordinate system is, naturally, where the *x*- and *y*-axes intersect. This point is designated as $P(0,0)$. The other interesting point shown in the figure is $P(x,y)$.

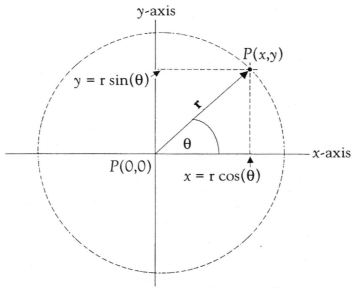

Figure 3-2: Polar and rectangular coordinates.

The point $P(x,y)$ can be thought of as a fixed point on the rim of a wheel. The axle is located at the point $P(0,0)$. The line from $P(0,0)$ to $P(x,y)$ is a *vector* specified as **r**. We can think of it as the radius of the wheel. (The variable **r** is shown in bold to indicate that it is either a vector or a complex variable.)

The variable **r** is often of interest in DSP, since its length is what defines the *amplitude* of the signal. This will become more clear shortly. When points are specified by their x and y values the notation is called *rectangular*. The point $P(x,y)$ can also be specified as being at the end of a line of length $\mathbf{r}$ at an angle of θ. This notation is called *polar* notation.

It is often necessary to convert between polar and rectangular coordinates. The following relationship can be found in any trigonometry book:

$$\text{length of } \mathbf{r} = \sqrt{x^2 + y^2} \qquad \text{Equation 3-18}$$

This is also called the magnitude of $\mathbf{r}$ and is denoted as $\left|\mathbf{r}\right|$. The angle θ is obtained from x and y as follows:

$$\theta = \arctan\left(\frac{y}{x}\right) \qquad \text{Equation 3-19}$$

Two particularly interesting relationships are:

$$x = \left|\mathbf{r}\right|\cos\theta \qquad \text{Equation 3-20}$$

and

$$y = \left|\mathbf{r}\right|\sin\theta \qquad \text{Equation 3-21}$$

The reason these two functions are so important is that they represent the signals we are usually interested in. In order to develop this statement further, it is necessary to realize that the system we have just described is static—in other words, the wheel is not spinning. In DSP, as with most other things, the more interesting situation occurs when the wheels start spinning.

From basic geometry, we know that the circumference of the wheel is simply $2\pi r$. This is important, since it defines the angular distance around the circle. If $\theta = 0$, then the point $P(x,y)$ will have a value of $P(|\mathbf{r}|,0)$. That is, the point will be located on the x-axis at a distance of $|\mathbf{r}|$ from the origin. As θ increases, the point will move along the dotted line. When $\theta = \pi/2$ the point will be at $P(0,|\mathbf{r}|)$. That is, it will be on the y-axis at a distance $|\mathbf{r}|$ from the origin. The point will continue to march around the circle as θ increases. When θ reaches a value of 2π, the point will have come full circle back to $P(|\mathbf{r}|,0)$.

As the point moves around the circle, the values of x and y will trace out the classic sine and cosine wave patterns. The two patterns are identical, with the exception that the sine lags the cosine by $\pi/2$. This is more often expressed in degrees of phase; the sine is said to lag the cosine wave by 90°.

When we talk about the point moving around the circle, we are really talking about the vector $\mathbf{r}$ rotating around the origin. This rotating vector is often called a *phasor*. As a matter of convenience, a new variable ω is often defined as:

$$\omega = 2\pi f \qquad \text{Equation 3-22}$$

The variable ω represents the *angular frequency*. The variable f is, of course, the frequency. Normally f is expressed in units of *hertz* (Hz), where 1 Hz is equal to 1 cycle per second. As we will see a little later, however, the concept of frequency can take on a somewhat surrealistic aspect when it is used in relation to DSP systems.

If all of this makes sense so far, you are in good shape with respect to the fundamentals of digital signal processing. If, however, all of this is a little hard to grasp, don't feel left out. Many engineers never really become completely comfortable with the mathematics. This isn't to say it's not important, however. The material in this section and the next *must* be well understood if you are to understand the mathematical principles of DSP.

The question is then: what should you do if this material seems vague? We have stepped through a lot of trigonometry quickly, so don't feel too bad if the material does not seem obvious. This section is intended only as a quick review. Also, the presentation in the book is naturally *static*, but phasors are a *dynamic* process. It is tough to get the feel of a dynamic process just by reading about it.

Interactive Exercise

We will talk about complex numbers next, but first it is worth noting that these relationships can be dynamically illustrated by graphing a complex exponential function. The program *cmplxgen* supplied on the accompanying disk is a good tool for this. To use it, just double-click on the icon. The program comes up with the appropriate values as a default. Then click on the GENERATE button. You can watch the point rotate and simultaneously see the waveforms that are generated for both the *x* and *y* values.

Feel free to change the values of amplitude and frequency. Adjust the frequency for values between 0.25 and 12. Adjust the amplitude for values between 0.25 and 1.25. Notice that it is OK to enter negative values, as long as they are in the same range. It may seem like an oversight that we have not included dimensions (like hertz or volts) on the above values. It isn't. This too will make sense as we proceed.

If, after working with *cmplxgen* for awhile, things still don't make sense, it is probably a good idea to find a basic book or study guide on trigonometry and do some studying. Then come back to this chapter for another try.

Complex Numbers

Now, back to the subject of complex numbers. We have stayed away from the subject until now simply because we did not want to confuse things. Partially because of the names used with complex numbers ("real" and "imaginary"), and partially because of their somewhat esoteric use, people are often intimidated by them. This is unfortunate, since complex numbers are really quite straight-forward. As with many other areas of mathematics, however, the notation can be a little confusing.

Part of the confusion over complex numbers—particularly as they relate to DSP—comes from a lack of understanding over their role in the "real world" (no pun intended). So, first we will present a qualitative discussion of real-world signals and complex numbers. After that, a more mathematical presentation will be in order. Complex numbers can be thought of as numbers with two parts: the first part is called the *real* part, and the second part is called the *imaginary* part. Naturally, most numbers we deal with in the real world are real numbers: 0, 3.3, 5.0, and 0.33 are all examples. Since complex numbers have two parts, it is possible to represent two related values with one number; *x-y* coordinates, speed and direction, or amplitude and phase can all be expressed directly or indirectly with complex numbers.

Initially, it is easy to think of signals as "real valued." These are what we see when we look at a signal on an oscilloscope, look at a time vs. amplitude plot, or think about things like radio waves. There are no "imaginary" channels on our TVs, after all.

In practice most of the signals we deal with are actually complex signals. For example, when we hear a glass drop we immediately get a sense of where the glass hit the floor. It is tempting to think of the signals hitting our ear as "real valued"—the amplitude of the sound wave reaching our ears as a function of time. This is actually an oversimplification, as the sound wave is really a complex signal. As the glass hits the floor the signal propagates *radially* out from the impact point. Imagine a stone dropped in a pond; its graph would actually be three-dimensional, just as the waves in a pond are three-dimensional. These three-dimensional waves are, in fact, complex waveforms. Not only is the waveform complex, but the signal processing is also complex. Our ears are on opposite sides of our head to allow us to hear things slightly out of phase. This phase information is perceived by our brains as directional information.

Another way to look at this is to compare a monaural system—such as an AM radio—with a stereo system. A good example of a stereo system is an FM radio. While stereo systems are so ubiquitous today that we take them for granted, at one time they were quite novel. The early stereos came with a demonstration record, typically a recording of a train. The sound would slowly start in the left speaker and then move across to the right speaker. The result was the sensation of hearing the train actually pass by. These demo records graphically illustrated the difference between complex and real-valued signals.

The brain can find the direction of an AM radio because it is processing the real signal as a complex waveform. The signal itself, however, is a point source. There is no way to tell which way a train is going if you hear it over a monaural (i.e., real) channel. In the case of a stereo signal, however, the brain processes a complex signal with complex detectors. Not only can the brain discern where the speakers are, but it can also tell which direction the train is moving.

The points we have been discussing, such as $P(0,0)$ and $P(x,y)$, are really complex numbers. That is, they define a point on a two-dimensional plane. We do not generally refer to them this way, however, as a matter of convention. Still, it is useful to remember that fact if things get too confusing when working with complex notation.

Historically, complex numbers were developed from examining the real number line. If we think of a real number as a point on the line, then the operation of multiplying by (-1) rotates the number $180°$ about the origin on the number line. For example, if the point is 7, then multiplying by (-1) gives us (-7). Multiplying by (-1) again rotates us back to the original value of 7. Thus, the quantity (-1) can be thought of as an operator that causes a $180°$ rotation. The quantity $(-1)^2$ is just one, so it represents a rotation of either $0°$, or equivalently, $360°$.

This leads us to an interesting question: If $(-1)^2 = 1$, then what is the meaning of $\sqrt{-1}$? There is no truly analytical way of answering the question. One way of looking at it, however, is like this: If 1 represents a rotation of $360°$, and (-1) represents a rotation of $180°$, then $\sqrt{-1}$ must, by analogy, represent a rotation of $90°$. In short, multiplying by $\sqrt{-1}$ rotates a value from the x-axis to the y-axis. Early mathematicians considered this operation a purely imaginary (that is, having no relation to the "real" world) exercise, so it was given the letter i as its symbol. Since i is reserved for current in electronics, most engineers use j as the symbol for $\sqrt{-1}$. This book follows the engineering convention.

In our earlier discussion, we pointed out that a point on the Cartesian coordinates can be expressed as $P(x,y)$. This means, in words, that the point P is located at the intersection of x units on the x-axis, and y units on the y-axis. We can use the j operator to say the same thing:

$$P(x,y) = P\left(|\mathbf{r}|\cos(\theta), |\mathbf{r}|\sin(\theta)\right)$$
$$= x + jy \qquad \textbf{Equation 3-23}$$

Thus, we see that there is nothing magical about complex numbers. They are just another way of expressing a point in the x-y plane. Equation 3-23 is important to remember since most programming languages do not support a native complex number data type, nor do most processors have the capability of dealing directly with complex number data types. Instead, most applications treat a complex variable as two real variables. By convention one is real, the other is imaginary. We will demonstrate this with some examples later.

In studying the idea of complex numbers, mathematicians discovered that raising a number to an imaginary exponent produced a periodic series. The famous mathematician Euler demonstrated that the natural logarithm base, e, raised to an imaginary exponent, was not only periodic, but that the following relationship was true:

$$e^{j\theta} = \cos\theta + j\sin\theta \qquad \textbf{Equation 3-24}$$

To demonstrate this relationship, we will need to draw on some earlier work. Earlier we pointed out that the sine and cosine functions could be expressed as a series:

$$\sin(x) = x - \frac{x^3}{3!} + \frac{x^5}{5!} - \dots$$ Equation 3-25

and

$$\cos(x) = 1 - \frac{x^2}{2!} + \frac{x^4}{4!} - \dots$$ Equation 3-26

Now, if we evaluate $e^{j\theta}$ using Equation 3-13 we get:

$$e^{j\theta} = 1 + j\theta - \frac{\theta^2}{2!} - \frac{j\theta^3}{3!} + \frac{\theta^4}{4!} - \frac{j\theta^5}{5!} - \frac{\theta^6}{6!} \dots$$

Equation 3-27

Expanding and rearranging Equation 3-27 gives us:

$$e^{-j\theta} = \sum_{m=0}^{\infty} \frac{(-1)^m \theta^{2m}}{(2m)!} + j \sum_{m=0}^{\infty} \frac{(-1)^m \theta^{2m+1}}{(2m+1)!}$$ Equation 3-28

Substituting Equation 3-25 and Equation 3-26 into Equation 3-28 gives us Equation 3-24.

Euler's relationship is used quite heavily throughout the field of signal processing, primarily because it greatly simplifies analytical calculations. It is much simpler to perform integration and differentiation using the natural logarithm or its base than it is to perform the same operation on the equivalent transcendental functions. Since this book is mainly aimed at practical applications, we will not be making heavy use of analytical operations using e. It is common in the literature, however, to use $e^{j\omega}$ as a shorthand notation for the common $\cos(\omega) + j\sin(\omega)$ expression. This convention will be followed in this book.

Euler's relationship can also be used as another way to express a complex number. For example:

$$P(x,y) = re^{j\theta}$$ **Equation 3-29**

is equivalent to Equation 3-23.

We have pushed the mechanical analogy about as far as we can, so it is time to briefly review what has been presented and then switch over to an electronic model for our discussion.

- The basic model of a signal is oscillatory motion.

- The simplest conceptualization is a point rotating about the origin.

- The motion of the point can be defined as:

$$P(x,y) = re^{j\omega}$$

  where $\omega = 2\pi f$, r is the radius, and f is the frequency of rotation.

- Euler's relationship gives us the following:

$$e^{j\theta} = \cos\theta + j\sin\theta$$
$$e^{-j\theta} = \cos\theta - j\sin\theta$$

The electronic equivalent of the wheel is the LC circuit. An example circuit is shown in Figure 3-3. By convention, the voltage is generally defined as the real value, and the current is defined as the imaginary value. The symbol ω is used to represent the resonant frequency and is determined by the value of the components. Assuming the resistance in the circuitry is zero, then:

$$e^{j\omega t} = \cos\omega t + j\sin\omega t$$ **Equation 3-30**

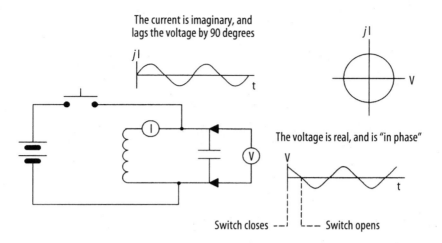

Figure 3-3: Ideal LC circuit showing voltage and current relationships.

describes the amplitude and the phase of the voltage and the current. In practice, we would add in a scale factor to define the value of the maximum voltage and the maximum current. Notice that, as in the case of the point rotating about the origin, the voltage is 90° out of phase with the current.

What if the resistance is *not* equal to zero? Then the amplitude decreases as a function of time. From any good book on circuit analysis, we can find that the decay of the amplitude is an exponential function of time: $e^{-\alpha t}$. This decay applies to both the current and the voltage. If we add in our scale factor A, we get the following equation:

$$f(t) = Ae^{-\alpha t}e^{j\omega t}$$

Equation 3-31

which, from our log identities, gives us:

$$f(t) = Ae^{(-\alpha + j\omega)t}$$

Equation 3-32

Generally, the exponential term is expressed as a single complex variable, s:

$$s = -\alpha + j\omega$$

Equation 3-33

The symbol s is familiar to engineers as the independent variable in the Laplace transform. (Transforms will be covered in a later chapter.)

Interactive Exercise

Now it's time to return to our program *cmplxgen*. In our previous example, we left the value of α at its default value of 0. Since $e^0 = 1$, this is equivalent to saying that the amplitude is constant, neither decaying nor increasing.

This time around, enter different values for the various options. Start out with the following:

$$\text{frequency} = 3$$
$$\text{amplitude} = 1.25$$
$$\alpha = -2$$

Notice that the resulting graph spirals in toward the origin. Try different values. Notice that positive values of α cause the graph to spiral out from the origin. Also notice that the amplitude of the sine waves changes as the point moves. This is a complex exponential at work!

This information on the complex exponential is critical to understanding how the major algorithms in DSP work, so make sure you feel comfortable with this material before proceeding.

A Practical Example

In order to illustrate some of the basic principles of working with discrete number sequences, we will begin with a simple example. Referring back to Figure 2-1, let's assume that our task is to use a DSP system to generate a sine wave of 1 Hz. We will also assume that our DAC has a resolution of 12 bits, and an output range of –5 volts to +5 volts.

This task would be difficult to do with conventional electronic circuits. Producing a sine wave generally requires an LC circuit or a special type of RC oscillator known as a Twin-T. In either case, finding a combination of values that work well and are stable at 1 Hz is difficult.

On the other hand, designing a low-frequency oscillator like this with DSP is quite straightforward. We'll take a somewhat convoluted path, however, so we can illustrate some important concepts along the way.

First, let's look at the basic function we are trying to produce:

$$f(t) = \sin(\omega t + \theta) \qquad \text{Equation 3-34}$$

where, for this example, $\omega = 2\pi f$, $f = 1$, and $\theta = 0$.

From a purely mathematical perspective, Equation 3-34 is seemingly simple. There are some interesting implications in this simple-looking expression, however. As Rorabaugh[1] points out, the notation $f(t)$ is used to mean different things by various authors. It may mean the entire function expressed over all values of t, or it may mean the value of f evaluated at some point t.

---

[1] *Digital Filter Designers Handbook*, page 36 (see References).

Another interesting concept is the idea that $f(t)$ is continuous. In practice, we know that no physical quantity is truly infinitely divisible. At some point quantum physics—if no other physical law —will define discretely quantized values. Mathematically, however, $f(t)$ is assumed to be *continuous, and therefore infinitely divisible*. That is, for any $f(t)$ and any $f(t + \Delta)$ there is some value equal to $f(t + \Delta/2)$. This leads to the rather interesting situation that between any two *finite* points on a line there are an *infinite* number of points.[2]

The object is to use a digital computer to produce an electrical output representing Equation 3-34. Clearly, we cannot compute an infinite number of points, as this would take an infinite length of time. We must choose some reasonable number of points to compute. What is a "reasonable number of points"? The answer depends on the system we are using and on how close an approximation we are willing to accept. In practice we will need something like 5 to 50 points per cycle. Figure 3-4 shows an example of how 16 points can be used to approximate the shape of a sine wave. Each point is called one *sample* of the sine function ($N = 15$).

Notice that time starts at $t = 0$ and proceeds through $t = {}^{15}/N$. In other words, there are 16 points, each evaluated at $1/16$-second intervals. This interval between samples is called (naturally enough) the sample period. The sample period is usually given the symbol T. Notice that the next cycle starts at $t = 0$ *of the second cycle*, so there is no point at the 1-second index mark. In order to incorporate T in an equation we must define a new term: the *digital frequency*. In our discussion of the basic trigonometry of a rotating point, we defined the angular frequency, ω, as being equal to $2\pi f$. The

---

[2] See pages 152–157 of *The Mathematical Experience* for a good discussion of this.

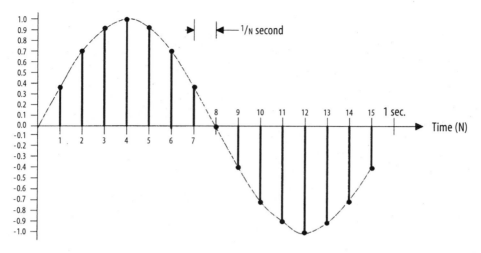

Figure 3-4: Sample points on a sine wave.

digital frequency λ is defined as the analog frequency times the period T:

$$\lambda = \omega T$$

$$= \frac{\omega}{N} \qquad \text{Equation 3-35}$$

The convention of using λ as the digital frequency is not universal. It was first used by Peled and Liu [2], and is used by Rorabaugh [3]. Giving the digital frequency its own symbol is useful as a means of emphasizing the difference between the digital and the analog frequencies, but is also a little confusing. In this text we denote the digital frequency as ωT. The justification for defining the digital frequency in this way will be made clear shortly.

The variable t is continuous, and therefore is not of much use to us in the computations. To actually compute a sequence of discrete values we have to define a new variable, n, as the index of the points. The following substitution can then be made:

$$t = nT, n = 0 \ldots N-1 \qquad \text{Equation 3-36}$$

Equation 3-35 and Equation 3-36 can be used to convert Equation 3-34 from continuous form to a discrete form. Since our frequency is 1 Hz, and there is no phase shift, the equation for generating the discrete values of the sine wave is then:

$$
\begin{aligned}
f(t) &= \sin\,(2\pi f t + \theta)\big|_n \\
&= \sin\,\big(2\pi(1)nT + 0\big), \ \ n = 0 \ldots N-1 \\
&= \sin\,(2\pi nT), \ \ n = 0 \ldots N-1 \qquad \text{Equation 3-37}
\end{aligned}
$$

Remember that T is defined as $1/N$. Therefore, Equation 3-37 is just evaluating the sine function at 0 to $^{N-1}/N$ discrete points. The need to include T in Equation 3-37 is the reason that the digital frequency was defined in Equation 3-35.

For a signal this slow, we could probably compute the value of each point in *real time*. That is, we could compute the values as we need them. In practice, however, it is far more efficient to compute all of the values ahead of time and then save them in memory. The first loop of the listing in Figure 3-5 is an example of a C program to do just this.

The first loop in Figure 3-5 generates the *floating point* values of the sine wave. The DAC, however, requires binary integer values to operate properly, so it is necessary to convert the values in k to properly formatted integers. Doing this requires that we know the binary format that the DAC uses, as there are a number of different types. For this example, we will assume that a 0 input to the DAC causes the DAC to assume its most negative (−5 V) value. A hexadecimal value of 0xFFF (that is, all ones) will cause the most positive output (+5 V).

```c
#include <stdio.h>
#include <math.h>

/* Define the number of samples. */
#define N 16

void main()
{

unsigned int DAC_values[N]; /* Values used by the DAC. */

double k[N]; /* Array to hold the floating point values. */
double pi; /* Value of pi. */

/* Declare an index variable. */
unsigned int n;

        pi = atan(1) * 4; /* Compute the value of pi. */

        for (n=0; n<N; n++)
                {
                k[n] = sin(2 * pi * ((float)n/(float)N));
                printf("%1.2f\n",k[n]);
                }

        for (n=0; n<N; n++)
                {
                DAC_values[n] = ((k[n] / 2.0) + 0.5) * 0xFFF;
                printf("%3X\n",DAC_values[n]);
                }

//    The following code is system dependent, so we have provided pseudo-
//    code to illustrate the types of things that need to be done. The
//    functions wait_seconds() and Output_to_DAC() are user defined.
//
//        while (1) /* Set up an infinite loop. */
//                {
//                for (n=0; n<N; n++)
//                        {
//                        wait_seconds (1/ (float) N); /* Wait 1/N seconds. */
//                        Output_to_DAC(DAC_values[n]); /* Output each value. */
//                        }
//                }
//
}
```

Figure 3-5: C listing for generating a sine wave.

The floating point values in $k[\ ]$ have a range of -1.0 to $+1.0$. The trick then is to convert these values so that -1.0 maps to 0x000 and $+1.0$ maps to 0xFFF. We can do this by dividing all of the values in k by 2, and then adding 0.5. This scales the values in k from 0.0 to 1.0. Then, we can multiply the values in k by 0xFFF. The result is a series of binary integers that represent equivalent values of the waveform. This operation is shown in the second loop of Figure 3-5.

The final step is to periodically (every $T = 1/N$ seconds) output the indexed value of $k[\ ]$. This step is highly system dependent, so it is not practical to present real code to perform the output function. At the bottom of Figure 3-5 is pseudocode that shows a typical sequence, however.

The result is shown in Figure 3-6. The stair-step shape is the output of the DAC. The dashed line is the ideal sine wave. After passing through the smoothing filter, the actual waveform will approximate the ideal.

This example is quite straightforward, but it does illustrate some very important concepts. One of these is, as we noted earlier, the concept of *digital frequency* vs. *analog frequency*. Previously we just defined the digital frequency as ωT, where T is equal to $1/N$ seconds, and N is the number of samples per seconds. In many practical applications, however, there is really no need to keep the relationship $T = 1/N$. For example, we can just assume that $T = 1$. Then, all we really care about is the ratio n/N; the value of T simply becomes a scaling factor. Another example will help illustrate the point.

In our previous example, we built a function generator, using digital techniques, to output a sine wave of 1 Hz. In that example, the digital and the analog frequency were the same thing. Now,

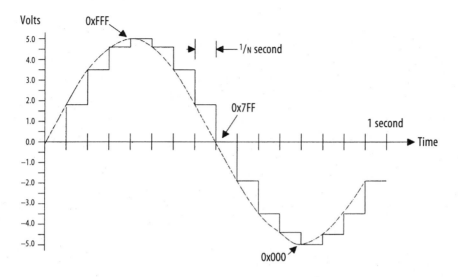

Figure 3-6: DAC output for a sine wave.

let's consider how to modify the output frequency of the function generator. There are actually two ways to accomplish this.

Let's assume we want to double the output frequency, from 1 Hz to 2 Hz. The first way to do this would be to *decrease the time* we wait to output the next sample to the DAC. For example, instead of waiting $1/N$ seconds to output the new value to the DAC, we could wait only $1/2N$ seconds to output the value. This would *double* the number of points that are output each second. Or, equivalently, we could think of this as outputting one cycle of the waveform in 0.5 seconds.

The important thing to notice here is that we have not re-evaluated Equation 3-37. We have changed the value of T but, as long as we understand what the implications are, there is no need to recompute the values of $f[n]$. The actual frequency output, interestingly enough, has *nothing* to do with the values computed. The actual (analog) frequency will match the digital (computed) frequency only when the output interval between points is equal to $1/N$ seconds. In this sense we see that digital frequency is computationally independent of the analog frequency.

This may seem a bit obtuse and esoteric, but it is of practical importance. Many DSP applications do not require real-time evaluation. For example, in seismic analysis the data is recorded first, and then processed. Processing a sample generally takes much longer than the time over which the signal was recorded. A 10-second record, for example, may take hours or days of computation time to process. In such situations, the value of T is critical only in scaling the final results. What counts computationally is the value N.

If this still seems a little fuzzy, don't feel too frustrated. For the moment, the key point we are trying to make is this: *In many DSP applications, the number of samples per some "unit period" determines how the signal is handled. Once processed, the signal is mapped back into real time by a scale factor T. T may or may not be directly related to $1/N$ seconds.*

What is the second way to change the output frequency? We could leave the output interval at $1/N$ seconds, and change the value of f in Equation 3-37. If we let $f = 2$, then Equation 3-37 becomes:

$$f(t) = \sin\left(2\pi ft + \theta\right)\big|_n$$
$$= \sin\left(2\pi(2)nT + 0\right),\ n = 0...N-1$$
$$= \sin\left(4\pi nT\right),\ n = 0...N-1$$
$$= \sin\left(\frac{4\pi n}{N}\right),\ n = 0...N-1 \qquad \textbf{Equation 3-38}$$

Notice that there will now be *two* cycles in 16 points. Each cycle of the sine wave will only have 8 points, as shown in Figure 3-7.

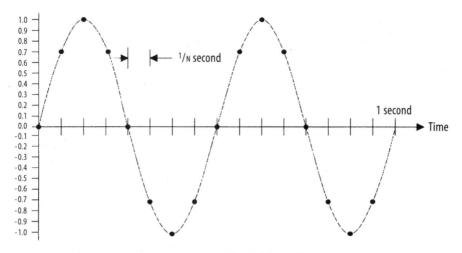

Figure 3-7: Two cycles of a sine wave.

This approach has the advantage that no adjustments have to be made to the output interval timing routines. On the other hand, the quality of the output waveform will vary as a function of frequency. This is because the number of points per cycle varies as a function of frequency. A practical DSP system must balance, and sometimes adjust in real time, the tradeoffs between the number of points used per second and the time interval between each point.

Chapter Summary

This chapter has discussed a number of mathematical relationships that are used extensively in digital signal processing. The emphasis has been on practical trigonometric relationships that are often overlooked in textbook discussions. This is particularly true concerning the role of complex numbers in trigonometric relationships.

In DSP, complex numbers are of practical importance: they are at the heart of many key DSP algorithms. There is, however, nothing magical about complex numbers. If we remember a couple of simple relationships, complex numbers can be handled as easily as any other number.

Finally, we introduced the concepts of analog and digital frequencies. The two are, of course, closely related. At the same time, they are strangely independent of each other. The analog frequency is often dropped in DSP calculations and the digital frequency used instead. Then, in the final result, the analog frequency is restored by scaling the digital frequency. Often this operation is left out in the discussion—a fact that can be very confusing.

In the following chapters, we'll apply the concepts developed here.

CHAPTER 4

Signal Acquisition

Introduction

In the last chapter we looked at ways to generate a signal using digital signal processing techniques. That discussion illustrated a number of key concepts that are fundamental to more sophisticated DSP applications. The concepts covered were the *number of samples per period*, the relationship of the sample *interval* to number of samples, and the related concept of analog vs. digital frequency. In this section we will carry the discussion further. We'll introduce the Nyquist theorem and discuss some practical considerations in choosing sampling rates.

In the previous chapter we produced signals from their mathematical definitions. This is an important and useful area of DSP known as *digital signal synthesis*. In most practical applications, however, we will be acquiring a signal and then doing some manipulation on this signal. This work is often called *digital signal analysis*.

One of the first things we must do when we are designing a system to handle a signal is to determine what performance is required. In other words, *how do we know that our system can handle the signal?* The answer to this question, naturally, involves a number of issues. Some of the issues are the same ones that we would deal with when designing any system:

- Are the voltages coming into our system within safe ranges?

- Will our design provide adequate bandwidth to handle the signal?

- Is there enough power to run the equipment?

- Is there enough room for the hardware?

We must also consider some additional requirements that are specific to DSP systems or are strongly influenced by the fact that the signals will be handled digitally. These include:

- How many samples per second will be required to handle the signal?

- How much resolution is required to process the signal accurately?

- How much of the signal will need to be kept in memory?

- How many operations must we do on each sample of the signal?

Stating the requirements in general terms is straightforward. We must ensure that the incoming analog signal is sufficiently bandwidth-limited for our system to handle it; the number of samples per second must be sufficient to accurately represent the analog signal in digital form; the resolution must be sufficient to ensure that the signal is not distorted beyond acceptable limits; and our system must be fast enough to do all required calculations.

Obviously, however, these are qualitative requirements. To determine these requirements explicitly requires both theoretical understanding and practical knowledge of how a DSP system works. In the next section we will look at one of the major design requirements: the number of samples per second.

Sampling Theory

In Equation 3-38 the frequency of the sine wave generated was increased by the value of the frequency *f*. This had the effect of increasing the number of cycles in a second—*at the cost of the number of samples per cycle*. In the example, there were 16 samples per second. Generating a frequency of 2 Hz meant that there were now only 8 samples per cycle. Similarly, if the frequency had been increased to 4 Hz, there would be only 4 samples per cycle.

The logical question is: How far can we carry the sequence? In other words, *what is the maximum frequency we can handle for a given number of samples per second?* We can get a good feeling for the answer by trying one more frequency: 8 Hz. Using the tools and techniques from Chapter 3 gives the graph shown in Figure 4-1. The dashed line is the expected analog signal. Notice, however, that all of the discrete points have a value of 0. We put a value of 8 into Equation 3-38, but we got out a DC value of zero. What went wrong?

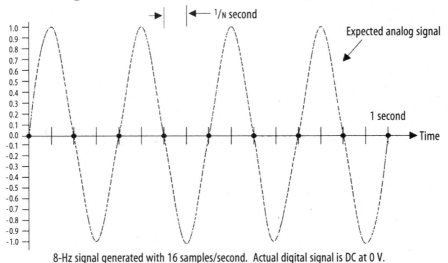

8-Hz signal generated with 16 samples/second. Actual digital signal is DC at 0 V.

Figure 4-1: Aliasing.

The answer to this question can be demonstrated for the general case when the frequency is equal to one-half the number of points. We can do this by plugging $f = N/2$ into Equation 3-38:

$$f(t) = \sin\left(2\pi f t + \theta\right)\big|_{t=nT}$$

$$= \sin\left(2\pi\left(\frac{N}{2}\right)nT + 0\right), \quad n = 0...N-1$$

$$= \sin\left(\pi n N T\right), \quad n = 0...N-1$$

$$= \sin\left(\frac{\pi n N}{N}\right), \quad n = 0...N-1$$

$$= \sin\left(\pi n\right), \quad n = 0...N-1 \qquad \text{Equation 4-1}$$

The sine function is 0 for a frequency of zero, *and* for integer multiples of π. We have therefore stumbled onto the answer to the question of what our maximum frequency is: *The frequency must be less than 1/2 the number of samples per second.* This is a key building block in what is known as the *Nyquist theorem*. We do not yet have all of the pieces to present a discussion of the Nyquist theorem, but we will shortly.

In the meantime, let's explore the significance of our discovery a little further. Clearly, this is another manifestation of the difference between the analog frequency and the digital frequency. Intuitively, we can think of it as follows: To represent one cycle of a sine wave, what are the minimum number of points needed? For most cases, any two points are adequate. If we know that any two separate points are points on *one* cycle of a sine wave, we can fit a curve to the sine wave. There is one important exception to this, however: when the two points have a value of zero. We need *more than* two points per cycle to ensure that we can accurately produce the desired waveform.

From the example above, we saw that we get the same output from Equation 4-1 if we put in a value for *f* of either 0 or 8 when we are using 16 samples/second. For this reason, these frequencies are said to be *aliases* of one another.

We just "proved," in a nonrigorous way, that our maximum digital frequency is $N/2$. But what happens if we were to put in values for *f* greater than $N/2$? For example, what if we put in a value of, say, 10 for *f* when N = 16? The answer is that it will alias to a value of 2, just as a value of 8 aliased to a value of 0. If we keep playing at this, we soon see that we can only generate output frequencies for a range of 0 to $N/2$.

Our digital frequency is defined as $\lambda = \omega T$. If we substitute $N/2$ for *f* and expand this we get:

$$\lambda = \omega T$$
$$= 2\pi f T$$
$$= 2\pi \left(\frac{N}{2} \right) \frac{1}{N}$$
$$= \pi \qquad \text{Equation 4-2}$$

It would therefore appear that our digital frequency must be between 0 and π. We can use any other value we want, but if it is outside this range, it will map to a frequency that is within the range of 0 to π. However, note that we said it would "*appear* that our digital frequency must be between 0 and π." This is because we haven't quite covered all of the bases.

Normally, in electronics we don't think of frequency as having a sign. As we saw in Chapter 2, however, negative frequencies are possible in the real world. Remember from that discussion that there is no great mystery to a negative frequency. It simply means that the phase between the real and imaginary components are opposite what they would be for a positive frequency. In the case of a point on the unit circle, a negative frequency means that the point is rotating clockwise rather than counterclockwise. The sign of the frequency for a purely real or a purely imaginary signal is meaningful only if there is some way to reference the phase.

The signals generated so far have been real, but there is no reason not to plug in a negative value of f. Since $\sin(-\omega) = -\sin(\omega)$, we would get the same frequency out, but it would be 180° out of phase. Still, this phase difference does make the signal unique; thus, the actual unique range of a digital frequency is $-\pi$ to π.

This discussion may seem a bit esoteric, but it definitely has practical significance. A common practice is to specify the performance of a DSP *algorithm* over the range of $-\pi$ to π. The DSP *system* will map this range to analog frequencies by selection of the number of samples per second.

The second part of demonstrating the Nyquist theorem lies in showing that what is true for sine waves will, if we are careful, apply to any waveform. We will do this in the section covering the Fourier series.

Sampling Resolution

In order to generate, capture, or reproduce a real-world analog signal, we must ensure that we represent the signal with sufficient resolution. Generally, resolution will have two characteristics:

- The number of samples per second.
- The resolution of the amplitude of each sample.

The resolution of the amplitude of each sample is a system parameter. In other words, it will depend upon the input circuitry, how the system is used, and so forth. However, the theoretical limit for the amplitude resolution is defined by the number of bits resolved in the ADC or converted by the DAC.

The formula for determining the resolution of a system is:

$$r_{min} = \frac{1}{2^n - 1} \qquad \text{Equation 4-3}$$

where n is the number of bits. For example, if we have a 2-bit system, then the maximum resolution will be:

$$r_{min} = \frac{1}{3}$$

Looking at this in table form shows the mapping for each of the possible binary values:

Binary Value	Weight
00	0
01	$1/3$
10	$2/3$
11	1

Notice that we have expressed the *weight* for each possible binary value. As with the case of digital versus analog frequency, we can only express the digital value as a dimensionless number. The actual amplitude depends on the scaling performed by the DAC or the ADC. Notice that in this example we are dealing with only positive values. In practice there are a number of different schemes for setting weights. Twos complement and offset binary are two of the most common schemes used in signal processing.

Let's look at a typical example. Assume that we are designing a system to monitor an important parameter in a control system. The signal has a possible range of −5 volts to +5 volts. Our analysis has shown us that we must know the voltage to within ±.05 volts. How many bits of resolution does our system need?

The first thing to do is to express the resolution as a ratio of the minimum value to the maximum range:

$$r_{min} = \frac{V_{min}}{V_{max}}$$

$$= \frac{0.05 \text{ volts}}{10 \text{ volts}}$$

$$= 0.005 \qquad \textbf{Equation 4-4}$$

We can now use Equation 4-3 to find the number of bits. In practice, we would probably try a couple of values of n until we found the right value. A more formal approach, however, would be to solve Equation 4-3 for n:

$$r_{min} = \frac{1}{2^n - 1}$$

$$2^n = \frac{1}{r_{min}} + 1$$

$$n = \log_2 \left(\frac{1}{r_{min}} + 1 \right) \qquad \textbf{Equation 4-5}$$

Plugging in 0.005 for r_{min} into Equation 4-5 yields a value for n of 7.651. Rounding this value up gives a value of eight bits. Therefore, we need to specify *at least* eight bits of resolution for our signal monitor. As a side note, most calculators do not have a $\log_2$ function. The following identity is handy for such situations:

$$\log_b(x) = \frac{\ln(x)}{\ln(b)}$$ Equation 4-6

In this example, we lightly skipped over the method for determining that we needed a resolution of 0.005 volts. Sometimes determining the resolution is straightforward, but sometimes it is not. As a general guide, you can make the following assumptions: Eight bits is adequate for coarse applications. This includes control applications that are not particularly sensitive, and signals that can tolerate a lot of distortion. Eight-bit resolution is adequate for low-grade speech applications, but twelve-bit resolution is much more common. This resolution is generally adequate for most instrumentation and control applications. Twelve-bit resolution produces telephone-quality speech. Sixteen-bit resolution is used for high-accuracy requirements. CD audio is recorded with 16-bit resolution. It turns out that 21 bits is about the maximum practical value for either an ADC or a DAC. Achieving this resolution is expensive, so 21-bit resolution is generally reserved for very demanding applications.

One final word is required on the subject of resolution in terms of the number of bits. The effect of quantizing a signal is to introduce noise. This noise is called, naturally enough, the quantization error. The noise can be thought of as the result of representing the smooth and continuous waveform with the stair-step shape of the digitally represented signal.

Chapter Summary

The performance of digital signal processing algorithms is generally specified by frequency response over a normalized frequency range of $-\pi$ to $+\pi$. The actual analog frequencies are scaled over this range by multiplying the digital frequency by the sample period. Accurately representing an analog signal in digital form requires that we convert from the digital domain to the analog domain (or the other way around) with sufficient resolution. In terms of the number of cycles, we must sample at a minimum of *greater than* twice the frequency of the sine wave. The resolution in terms of the amplitude depends upon the application.

CHAPTER 5

Some Example Applications

Introduction

At this point let's take a look at where we have been and where we are going. So far, we've been concerned with the mechanics of getting a signal into and out of our DSP system, and with reviewing some general math principles we will use later on. We have seen that we can sample a waveform, optionally store it, and then send it back out to the world. This is, in and of itself, a very useful ability. However, it represents only a small fraction of the things we can do with a DSP system.

The rest of this book will be taken up with examining the *other* things that we can do. Understanding how a DSP system is designed and used basically requires two types of knowledge. The first is an understanding of the applications that lend themselves best to DSP. The second type is an understanding of the tools necessary to design the system to accommodate these applications.

Most DSP texts, and even most engineering courses, focus only on the tools necessary for designing DSP algorithms. Often, there is little or no emphasis on *why* these tools are important, where they would be required, what practical utility they bring to the process, or how to start a design from a blank piece of paper.

This is unfortunate for a couple of reasons. One is that it leaves the student to trust that the mathematical discussions of the techniques will, sometime in the future, be of some practical use. Most students who have tried to bring purely academic training to bear on real-world design problems are justifiably suspicious of this assumption. It reduces the motivation to understand the material and contributes to much of the frustration many students find in studying DSP techniques.

However, we have a more immediate and practical reason for not liking this approach. Many of the key concepts in DSP are understood in terms of other DSP concepts. What this means in practice is that there is a *critical mass* of knowledge required for a basic understanding of the DSP techniques. In my experience, it is much easier to understand how the techniques and tools fit together if they are presented in reference to real applications. This provides guidance as to why a particular technique is required, helps to tie the techniques together in a common framework, and removes much of the abstraction from the process.

With this in mind, let's now turn our attention to the subject of filtering, beginning with a simple filter that is easily understood intuitively. We will then move on to developing the tools and techniques that will allow us to create more sophisticated, higher-performance filters of professional quality.

Filters

One of the most common DSP operations is filtering. As with analog filters, DSP filters can provide low-pass, bandpass, and high-pass filtering. (Specialized functions, such as notch filters, are also possible, though we will not be covering them in this book.)

The basic idea behind filtering in general is this: An input signal, generally a function of time, is input to a *transfer function*. Normally, the transfer function is a differential equation expressed as a function of frequency. The output of the transfer function is some *subset* of the input signal.

A block diagram of a low-pass filter is shown in Figure 5-1. In the figure, the input signal is a sum of two sine waves: one of them at a fundamental frequency, the other at the third harmonic. After passing through the transfer function $H(\omega)$ only the fundamental frequency remains; the first harmonic has been blocked. The top portion of Figure 5-1 depicts the low-pass filter as a function of time. The bottom portion of Figure 5-1 shows the filter as a function of frequency. We will be revisiting these concepts in greater detail in later chapters.

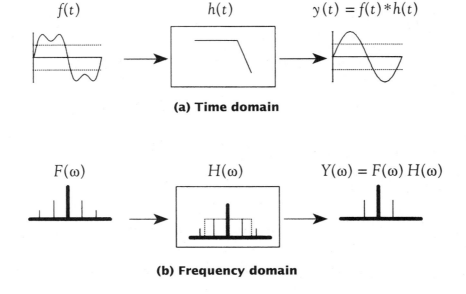

$f(t)$ $h(t)$ $y(t) = f(t) * h(t)$

(a) Time domain

$F(\omega)$ $H(\omega)$ $Y(\omega) = F(\omega)\,H(\omega)$

(b) Frequency domain

Figure 5-1: The basic low-pass filter.

In the world of analog electronics, the transfer function $H(\omega)$ is realized by arranging a combination of resistors, capacitors, inductors, and possibly operational amplifiers. In DSP applications, a computer is substituted for the resistors, capacitors, and inductors. The computer then computes the output using the input and $H(\omega)$.

The question for the DSP applications developer then becomes: How do we define $H(\omega)$ to give us the desired transfer function? This chapter shows, in an intuitive way, how simple digital filters operate. After that, several key concepts are introduced that lay the groundwork for developing more sophisticated filters. In the next chapters, we will see how to apply these tools to develop some practical working filters.

A Simple Filter

First, let's examine a simple application. Consider, for example, that much of the most interesting music of the twentieth century is stored on phonograph records. These records store their data using variations in the groove running from the outside of the record to its center. Over time, peaks in the groove can break off, or dents can be forced in the walls of the groove. When the phonograph needle hits one of these obstructions, the result is a "pop" in the music being played, as shown graphically in Figure 5-2. A pop is shown riding on an otherwise clean sine wave.

As these records are converted to CDs or tapes, it is natural to look for ways to eliminate these pops, thus restoring the more natural sound of the recording. One obvious solution is to manually adjust the spike down to a level where it is consistent with the rest of the signal. This could be done with a waveform editor or, in this simple case, even with a spreadsheet program.

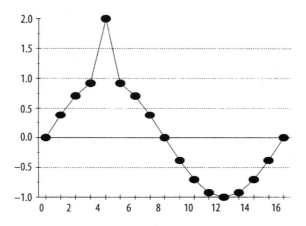

Figure 5-2: A noise "pop" on a sine wave.

Actually, manually editing the waveform is a good approach since it makes use of the best signal processor in the world: the human brain. For critical passages, it is fairly common for a person to manually edit the waveform. However, this approach is quite labor intensive. CDs are sampled at 44 kHz, and manually searching 44,000 points for each second of music rapidly becomes prohibitive. It's reasonable to find a more automated approach.

One simple approach is to average the value on either side of the spike with the value of the spike. This would not eliminate the spike, but it certainly would minimize it. We can do this using a simple algorithm:

$$g(n) = \frac{f(n-1) + f(n) + f(n+1)}{3}$$

Equation 5-1

Table 5-1 shows what happens when we apply this averaging routine to the signal in Figure 5-2. Notice that we have applied the averager across the entire signal from $n = -1$ to $n = 17$. This has the effect of moving the center point along the waveform. Therefore, this type of filter is known as a *moving average* filter. Notice that the table actually starts before the first sample—that is, we start evaluating $g(n)$ for $n = -1$. This might seem a little strange, but it makes sense when you consider that one of the terms in $g(n)$ is $f(n + 1)$. By starting at $n = -1$, we begin evaluating the signal at $f(0)$. For the first output value that we compute, $n = -1$, we have defined $f(-2)$ and $f(-1)$ to be zero. In a similar fashion, the value of $f(n + 1)$ is defined to be zero when $n = 16$ and $n = 17$.

The averaged values closely track the original values except at $n = 4$. For $n = 4$ the average value is much smaller than the input value. It is, in fact, much closer to where we want it. This routine does a fairly good job of minimizing the pops in a recording. Figure 5-3 is a graph of the original function and the output of our averaging routine.

Let's look more closely at how and why this routine works. Most of the changes in values

Table 5-1

n	$f(n)$	$\dfrac{f(n-1) + f(n) + f(n+1)}{3}$
−1	0.000	0.000
0	0.000	0.128
1	0.383	0.363
2	0.707	0.671
3	0.924	1.210
4	2.000	1.283
5	0.924	1.210
6	0.707	0.671
7	0.383	0.363
8	0.000	0.000
9	−0.383	−0.363
10	−0.707	−0.671
11	−0.924	−0.877
12	−1.000	−0.949
13	−0.924	−0.877
14	−0.707	−0.671
15	−0.383	−0.363
16	0.000	−0.128
17	0.000	0.000

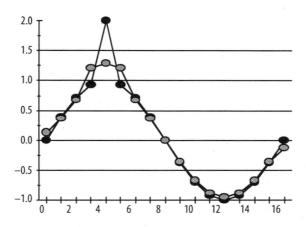

Figure 5-3: Effects of a moving average filter.

from one point to the next point in the original signal are relatively small. Therefore, for most points, the average value of the three points is relatively close to the center value. At $n = 4$ in the original signal, however, the value makes a large (or, equivalently, *rapid*) change. The moving average routine prevents this rapid change from propagating through.

In summary, the action of the averager has little effect on slowly changing signals and a much larger effect on rapidly changing signals. This is equivalent to saying that low-frequency signals suffer little attenuation, while high-frequency signals are strongly attenuated. That is, of course, the definition of a *low-pass filter*.

While it is clear that Equation 5-1 represents a low-pass filter, it is *not* clear exactly what the frequency response of the filter is. One conceptually simple way to find the frequency response of this filter is to measure the response for a variety of sinusoidal inputs. For example, let's divide the frequencies between 0 and π into six frequencies. Next, feed into the filter cosine waveforms at these frequencies and measure the peak output. We picked a *cosine* waveform because it gives us a value of 1 for an input of 0 Hz, keeping the response consistent with a low-pass filter.[1] With this information, we can then create a table of the frequency response, as shown in Table 5-2. From this table we can graph the frequency response of our low-pass filter; the graph is shown in Figure 5-4.

So far our development of the low-pass filter, and its response, has been very empirical. This is often how it is done in the real world. For example, the financial community often makes use of moving averages to filter out the day-to-day variations in stock prices, commodity prices, etc. This filter allows the stock analysts

Table 5-2

Frequency (cosine wave input)	Response (peak amplitude)
0.000	1.000
$\pi/5$	0.873
$2\pi/5$	0.539
$3\pi/5$	0.373
$4\pi/5$	0.167
π	0.000

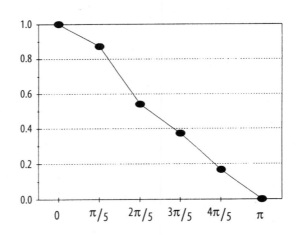

Figure 5-4: Frequency response of a simple filter.

---

[1] A sine function with an input of 0 Hz produces an output of 0. A cosine function with an input of 0 Hz produces an output of 1. Had we used a *sine* wave, the 0 Hz input value would have produced an output value of 0. This is mathematically acceptable, but it would not be consistent with generating test data for a low-pass filter. In this respect, a sine wave of 0 Hz is a bit anomalous. This situation of switching between a sine and a cosine wave is a fairly common trick in the literature.

to see the underlying trend of the price, without having the trend line distorted by transient perturbations.

On the other hand, this empirical approach can be difficult to manage for more sophisticated filters. As can be seen from Figure 5-4, the moving average filter is not very "crisp." It gradually attenuates the signal as the frequency increases. Often, we are more interested in a "brick wall" filter, which is a filter that does not affect the signal at all up to a cutoff frequency, then reduces any higher frequency components to zero above the cutoff.

Shortly we will look at more formal ways of developing and evaluating filters. But first let's explore these intuitive filters a little more.

Let's revisit Figure 5-2. On our last pass the signal was the sine wave and the noise was the spike. It could just as easily have been the other way around, however. For example, one problem that constantly plagues engineers is the presence of the 60-Hz "hum" created by the ubiquitous AC power wiring. This problem generally manifests itself as a sine wave superimposed on top of the signal of interest. A typical example is a system that monitors photons. When a photon strikes a detector, it produces a small electrical pulse. The result of such a pulse on top of the 60-Hz hum would look like Figure 5-2.

How can we eliminate the 60-Hz hum and leave the signal relatively intact? Obviously, our moving average filter will not do the job in this case. It does, however, suggest a solution. If we took the average of the points, and then *subtracted* this average value from the center value, we get the desired result. Algorithmically:

$$g(n) = f(n) - \frac{f(n-1) + f(n) + f(n+1)}{3} \qquad \text{Equation 5-2}$$

Table 5-3 shows the results of applying Equation 5-2 to the data shown in Figure 5-2. The graphical result is shown in Figure 5-5. Notice that the sine wave is essentially eliminated, leaving only the spike. Just as the moving average filter represented a low-pass filter, this differential filter represents a *high-pass* filter; the low-frequency

Table 5-3

n	$f(n)$	$f(n)-f(n-1)+f(n)+f(n+1)$
−1	0.000	0.000
0	0.000	−0.128
1	0.383	0.019
2	0.707	0.036
3	0.924	−0.286
4	2.000	0.717
5	0.924	−0.286
6	0.707	0.036
7	0.383	0.019
8	0.000	0.000
9	−0.383	−0.019
10	−0.707	−0.036
11	−0.924	−0.047
12	−1.000	−0.051
13	−0.924	−0.047
14	−0.707	−0.036
15	−0.383	−0.019
16	0.000	0.128
17	0.000	0.000
18	0.000	0.000

sine wave is heavily
attenuated, the high-
frequency spike
is only moderately
attenuated.

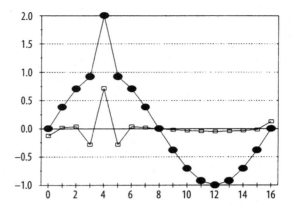

**Figure 5-5: Effects of a
difference filter.**

These two ex-
amples illustrate in
an intuitive way how
digital filters work.
In practice, most
common digital filters
are simply more sophisticated
versions of these simple filters. A bandpass filter, for example, can
be achieved by combining a low-pass filter and a high-pass filter.

Causality

Causality refers to whether a filter can be implemented in *real
time*. This is not a very strong definition of causality, but we do not
yet have the mathematical tools to define the term more precisely.

What does causality mean from an intuitive standpoint? We
can get a good idea by looking back at our moving average filter.
Notice that for any given sample n, we used both $n - 1$ and $n + 1$
sample points as well. If we think about n as being the current
sample (that is, the one coming in immediately), we obviously
have a problem. Getting the $n + 1$ sample means that we must
know the *future* value of f.

In our recording example, this was not a problem. Since the
data is recorded, we can find values for points that appear, with
respect to n, to be both in the future ($n + 1$) and in the past ($n - 1$).

For a real-time application, however, this is not an option; we are constrained to using only the current and past values of f. Filters that require only current and past values of a signal are called *causal filters*. Filters, such as our moving average filter, that require future values are called *noncausal* filters. As a matter of perspective, all real-world analog filters are causal. This is another example of the advantage of DSP: it allows us to build filters that could not be realized in any other way.

Notice that we can make our filter causal by simply shifting the index back by one:

$$y(n-1) = \frac{f(n+1) + f(n) + f(n-1)}{3}$$ **Equation 5-3**

which is equivalent to:

$$y(n) = \frac{f(n) + f(n-1) + f(n-2)}{3}$$ **Equation 5-4**

Equation 5-4 will not work quite as well as the noncausal version, since it is not symmetrical about the sample point. It will work nearly as well, however. In fact, the difference may be virtually undetectable in many applications. More important for our discussion is the fact that it does not significantly change our *conceptualization* of how the moving average filter works.

Convolution

Convolution is one of the key concepts in DSP. In its simplest terms, convolution is the process of feeding one function into (or as it is sometimes called, *through*) another function. Conceptually, for example, a filter can be thought of as a function. When we feed some function (such as the one in Figure 5-2) through our moving average filter, we are *convolving* the input function with the moving average filter. The asterisk (*) is normally used to denote convolution:

$$y[n] = f[n] * h[n]$$

Equation 5-5

where $h[n]$ are the coefficients of our filter, and $f[n]$ is the input function. In our moving average filter $h[n]$ had three coefficients and they were all equal to $1/3$.

Convolution is sufficiently important that it is worth developing the subject in detail. In the following examples, the notation will be somewhat simplified. Instead of using $f[n]$, we will use the simpler f_n. The meaning is identical.

In review then, our moving average filter can be expressed as follows:

$$y[n] = \frac{f[n+1] + f[n] + f[n-1]}{3}$$

Equation 5-6

Distributing the $(1/3)$ gives us:

$$y[n] = \frac{1}{3}f[n+1] + \frac{1}{3}f[n] + \frac{1}{3}f[n-1]$$

Equation 5-7

To make the expression more general, we replace the constants with the function h:

$$y[n] = h_0 f[n + 1] + h_1 f[n] + h_2 f[n - 1]$$

Equation 5-8

Converting to our simpler notation yields:

$$y_n = h_0 f_{n+1} + h_1 f_n + h_2 f_{n-1}$$

Equation 5-9

It is worthwhile to study the actual computation sequence that goes on in the filter. Let's take the first four samples of f: f_0, f_1, f_2 and f_3.

We start out at time $n = -1$. The first computation is then:

$$y_{-1} = h_0 f_0 + h_1 f_{-1} + h_2 f_{-2} \qquad \textbf{Equation 5-10}$$

Immediately, a problem crops up. We require values of f with a negative index. In other words, we need values *before* our first sample. We can get around this problem by simply defining f to be 0 at any point where it is not explicitly defined. Thus, for $n = -1$ we obtain:

$$y_{-1} = h_0 f_0 \qquad \textbf{Equation 5-11}$$

This notation is still a little awkward, since the y_{-1} implies that our first output occurs at some time prior to the $n = 0$ point. This is just a manifestation of our noncausal implementation. It really is our *first* output.

In a similar fashion, we can get the next output for $n = 0$:

$$\begin{aligned} y_0 &= h_0 f_1 + h_1 f_0 + h_2 f_{-1} \\ &= h_0 f_1 + h_1 f_0 \qquad \textbf{Equation 5-12} \end{aligned}$$

Proceeding along these lines, we obtain the results shown in Table 5-4. Notice the symmetry and pattern of the terms in the table. We have been careful to line up the terms in the equations to emphasize this point. With a little contemplation, we can derive a very compact expression for producing the terms in Table 5-4:

$$y[n] = \sum_{k=-\infty}^{\infty} h[k]\, f[n-k] \qquad \textbf{Equation 5-13}$$

Table 5-4

$$y[-1] = h_0 f_0$$

$$y[0] = h_0 f_1 + h_1 f_0$$

$$y[1] = h_0 f_2 + h_1 f_1 + h_2 f_0$$

$$y[2] = \qquad h_1 f_2 + h_2 f_1 + h_3 f_0$$

$$y[3] = \qquad\qquad h_2 f_2 + h_3 f_1$$

$$y[4] = \qquad\qquad\qquad h_3 f_2$$

One caveat: Don't try to apply Equation 5-13 too literally to produce Table 5-4, as the $n = -1$ term will throw you off. If you start with $n = 0$, however, you will get the same *terms* shown in Table 5-4. More formally, we can say that Equation 5-13 is valid for all non-negative index values of y.

Equation 5-13 is called the *convolution sum*, and we can use it directly to implement filters. We simply plug in the coefficients for h, and then feed in the values for the input function f. Obviously, finding the coefficients for h is of key interest. So far we have only been able to come up with the simple moving average filter:

$$h[n] = \frac{1}{N}, n = 0,\ 1,\ 2 \dots N - 1 \qquad \text{Equation 5-14}$$

Increasing N gives more terms to average, and therefore a lower frequency response. Fewer terms give fewer terms to average, and therefore a higher frequency response. As we saw, we can empirically determine the curve for the frequency response, but we cannot really do much to control the *shape* of the curve.

It would be much more useful if we could simply draw the

frequency response we wanted, and then convert that frequency response to the coefficients for h. That is exactly what we will do, but first we must develop a few more tools.

Chapter Summary

In this chapter we accomplished two things. First, we demonstrated how a low-pass filter and a high-pass filter can be developed from a heuristic standpoint. Next, we presented one of the basic concepts needed to develop more sophisticated filters: *convolution*.

As is generally the case in mathematics, it is difficult to appreciate abstract concepts like convolution of discrete sequences without seeing some practical application. If these concepts don't make sense at this point, don't worry. As long as you got the general idea, you will be prepared for the work ahead. If the material is unclear at this point, we recommend reading ahead and looking at how these tools are applied. Then, if necessary, come back and reread this chapter. It will make more sense then.

CHAPTER 6

The Fourier Series

Introduction

In this chapter we will be discussing the Fourier series. The Fourier series plays an important theoretical role in many areas of DSP. However, it generally does not play much of a practical role in actual DSP system design. For this reason, we will spend most of this section discussing the insights to be gained from the Fourier series; we will not devote a great deal of time to the mathematical manipulations commonly found in academic texts.

Background

The Fourier series is named after the French mathematician Joseph Fourier. Fourier and a number of his contemporaries were interested in the study of vibrating strings. In the simple case of just one naturally vibrating string the analysis is quite straightforward: the vibration is described by a sine wave. However, musical instruments, such as a piano, are made of many strings all vibrating at once. The question that intrigued Fourier was: How do you evaluate the waveforms from a number of strings all vibrating at once?

As a product of his research, Fourier realized that the sound heard by the ear is actually the arithmetic sum of each of the individual waveforms. This is called the *principle of superposition*. This is not such a dramatic observation and is, in fact, somewhat intuitive.

The really interesting thing that Fourier contributed, however, was the realization that virtually any physical waveform can, in fact, be represented as the sum of a series of sine waves.

The Fourier Series

Figure 6-1 shows an example of how the Fourier series can be used to generate a square wave. The square wave can be approximated by the expression:

$$f(t) = \sin \omega t + \frac{1}{n} \sin (n\omega t), \, n = 1, 3, 5, 7, \ldots, \, \infty \qquad \text{Equation 6-1}$$

The first term on the right side of Equation 6-1 is called the fundamental frequency. Each value of n is a *harmonic* of the fundamental frequency.

Looking at Figure 6-1, we can see that after only two terms the waveform begins to take on the shape of a square wave. Adding in the third harmonic produces a closer approximation to a square wave. If we keep adding in harmonics, we continue to obtain a waveform that looks more and more like a square wave. Interestingly enough, even if we added an infinity of odd harmonics we would not get a perfect waveform. There would always be a small amount of "ringing" at the edges. This is called the Gibbs phenomena.

There are some very interesting implications to all of this. The first is the fact that the bandwidth of a signal is a function of the *shape* of a waveform. For example, we could transmit a 1-kHz sine wave over a channel having a bandwidth of 1 kHz, but if we wanted to transmit a 1-kHz *square* wave we would have a problem.

Equation 6-1 tells us that we need infinite bandwidth to transmit a square wave! And, indeed, to transmit a *perfect* square wave would require infinite bandwidth. However, a perfect square

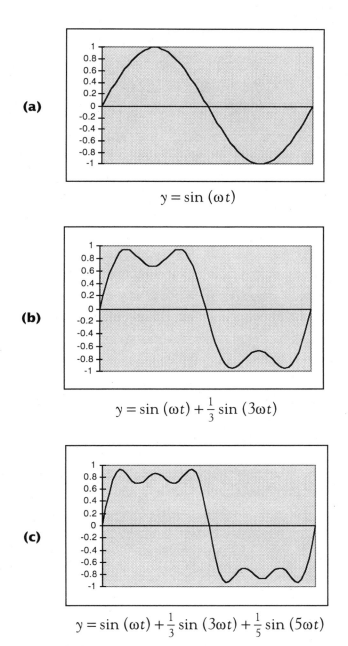

(a)

$$y = \sin(\omega t)$$

(b)

$$y = \sin(\omega t) + \frac{1}{3}\sin(3\omega t)$$

(c)

$$y = \sin(\omega t) + \frac{1}{3}\sin(3\omega t) + \frac{1}{5}\sin(5\omega t)$$

**Figure 6-1: Creating a square wave
from a series of sine waves.**

wave is *discontinuous;* the change from the low state to the high state occurs in zero time. *Any* physical system will require *some* time to change state. Therefore, any attempt to transmit a square wave must involve a compromise.

In practice, 10 to 15 times the fundamental frequency provides enough bandwidth to transmit a high-quality square wave. Thus, to transmit our 1-kHz square wave would require something like a 10-kHz bandwidth channel. A wider channel would give a sharper signal, while a narrower channel would give a more rounded square wave.

These observations lead to some interesting correlations. The higher the frequency that a system can handle, the faster it can change value. Naturally, the converse is true: The faster a system can respond, the higher the frequency it can handle.

This information also gives us the tools to complete the development of the Nyquist theorem.

The Nyquist Theorem Completed

In Chapter 4 we demonstrated that we needed at least two non-zero points to reproduce a sine wave. This is a necessary but not sufficient condition. For any two (or more) non-zero points that lie on the curve of a sine wave, there are an infinite number of harmonics of the sine wave that will also fit the same points. We eliminated the harmonic problem by requiring that all of our samples be restricted to one cycle of the sine wave. We will revisit this limitation in a minute, but first let's look closer at our work on the Nyquist theorem up to this point.

The big limitation on our development of the Nyquist theorem so far has been the requirement that we only deal with sine waves.

By taking into account the Fourier series we can remove this limitation. The Fourier series tells us that, for any practical waveform, we can think of it as the sum of a number of sine waves. All we need to concern ourselves with is handling the highest frequency *present* in our signal.[1] This allows us to state the Nyquist theorem in the form normally seen in the literature:

> *To accurately reproduce a signal, we must sample at a rate **greater than twice** the frequency of the highest frequency component **present** in the signal.*

The bold emphasis is to highlight two areas that are often misinterpreted. It is often stated that it is necessary to sample at *twice* the highest frequency of interest. As we saw earlier, sampling at twice the frequency only guarantees that we will get two points over one cycle. If these two points occur at the zero crossing, it would be impossible to fit a curve to the two points.

Another common mistake is to assume that it is sufficient to sample a signal at twice the frequency *of interest*. It is not the frequency of interest, but rather the frequency *present* that is important. If there are signal components higher in frequency than the Nyquist frequency, they will be aliased into the frequency below the Nyquist frequency and cause distortion of the sampled signal.

The next logical question then is: How do we ensure that aliasing does not occur? The solution to this problem brings us back to the anti-aliasing filter. In theory, we set the cutoff frequency of the anti-aliasing filter just below the Nyquist frequency. This

---

[1] To be more precise, this is strictly true only for *base-band* (that is, unmodulated) signals. We can, in fact, exploit aliasing to demodulate a signal using a technique called *sub-sampling*. Sub-sampling is beyond the scope of this book.

ensures that no frequency components equal to or greater than the Nyquist frequency can be sampled by the rest of the system, and therefore no aliasing of signals can occur. This removes our earlier restriction that the two points be located on one cycle of the waveform. The anti-aliasing filter ensures that this case is met for the highest frequency. In practice, we seldom try to push the Nyquist frequency. Generally, instead of sampling at twice the frequency, we will sample at five to ten times the highest frequency we are trying to capture.

This is easiest to demonstrate with an example. Let's say that we are interested in building a DSP system that can record voices at telephone-quality levels. Generally, telephone-quality speech can be assumed to have a bandwidth of 5 kHz. Even though the human hearing range is generally defined as 20 Hz to 20 kHz, most speech information is contained in the spectrum below 5 kHz.

The limiting factor on an analog voice input is generally the microphone. These typically handle frequencies up to 20 or 30 kHz, though the cheaper mikes will start rolling off in amplitude around 10 kHz or so. Thus, there will be frequency components present that are well above our upper frequency of interest. An anti-aliasing filter is needed to eliminate these components.

If we assume that we want to sample our signal at five times the highest frequency of interest, then our sampling rate would be 25 kHz. Strictly speaking, this would dictate a Nyquist frequency of 12.5 kHz. However, since we are not interested in frequencies this high, it makes sense to set the cutoff of the anti-aliasing filter at around 6 kHz or so. This gives us some headroom above our design requirement of 5 kHz, but is low enough that we will be *oversampling* the signal by a factor greater or equal to $12.5\text{ kHz}/6\text{ kHz}$. This oversampling allows us to relax the performance specifications on the analog parts of the system, thus making our system more robust and easier to build.

Setting the cutoff of the anti-aliasing filter well below the Nyquist frequency has another significant advantage: it allows us to specify a simpler filter with a slower roll-off. Such a filter is cheaper and introduces much less phase distortion.

Chapter Summary

The Fourier series tells us that any practical signal can be represented as a series of sine waves. This allows us to do all of our analysis of systems using only sinusoidal inputs—a very significant simplification! By looking at the harmonics of any signals that we wish to understand, we can gain a good understanding of the bandwidth requirements for our system. This analysis allows us to specify the sampling rate and the practical frequency cutoffs necessary to implement a practical system.

CHAPTER 7

Orthogonality and Quadrature

Introduction

The study of DSP can be confusing and frustrating. Hopefully, the material presented so far has been sufficiently clear to help alleviate some of this confusion and frustration. One of the areas that is often subject to confusion is the concept of orthogonality. Most DSP textbooks will at least mention the concept, but few actually explain it thoroughly. This is unfortunate, since orthogonality is one of the basic building blocks upon which all DSP work is based. Without a good understanding of orthogonality, many DSP concepts are nearly impossible to grasp at the intuitive level.

This subject is not particularly complicated. However, since it is so critical to an understanding of DSP, we will take some time to develop it in detail.

Orthogonality

The term *orthogonality* derives from the study of vectors. Most likely you have run across the term in basic math courses on trigonometry or calculus. By definition, two vectors in a plane are orthogonal when they are at a 90° angle to each other. When this is the case, the dot product of two vectors is equal to zero:

$$\xrightarrow[v_1]{} \bullet \uparrow v_2 = 0$$

The main point here is that the idea of multiplying two things together and getting a result of zero has been generalized in mathematics under the term *orthogonality*.

We will get back to this shortly, but let's look at another case where an interesting function has a zero value: the average value of a sine wave. Figure 7-1 shows one cycle of a sine wave. We have shaded in the area under the curve for the positive cycle and the area above the curve for the negative cycle. Notice that the area for the negative portion of the waveform is labeled with a negative symbol. A "negative area" is a hard concept to imagine, but be reassured that we are simply talking about an area that has a negative sign in front of it.

If we add the two areas together we will, naturally, get a value of zero. This may seem too obvious to bother pointing out, but it is just the first step. As an interesting side note, this fact was used in the early days of electricity to "prove" that AC voltages were of no practical use. Since they averaged to zero, so the analysis went, they could not do work!

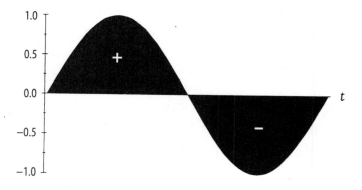

Figure 7-1: The average area under a sine wave is zero.

The process of integration can be viewed as finding the area under a curve. Therefore, you can write this idea mathematically as follows, for any integer value of k:

$$\int_0^{2\pi k} \sin \omega t \, dt = 0 \qquad \text{Equation 7-1}$$

Now, if you multiply by a constant, on both sides of the integral, the result is still the same:

$$\int_0^{2\pi k} A \sin \omega t \, dt = A \int_0^{2\pi k} \sin \omega t \, dt = 0 \qquad \text{Equation 7-2}$$

That is, the amplitude of the waveform may be larger or smaller, but the average value is still zero.

Now we come to the interesting part. What if we put in, not a constant, but some function of time? That is:

$$\int_0^{2\pi k} g(t) \sin \omega t \, dt = ? \qquad \text{Equation 7-3}$$

The answer naturally depends upon what our function of $g(t)$ is. But as we saw in the last chapter, we really only need to worry about sinusoidal functions for $g(t)$. We can extend our analysis to other waveforms by simply considering the Fourier representation of the waveform. Let's look at the specific case where $g(t) = \sin \eta t$.

$$\int_0^{2\pi k} \sin \eta t \sin \omega t = 0, \; \eta \neq \omega \qquad \text{Equation 7-4}$$

Equation 7-4 is called the *orthogonality of sines*. It tells us that, as long as the two sinusoids do not have the same frequency, then the integral of their products will be equal to zero. This may be a little hard to visualize. If so, think back to Equation 7-2. When the frequencies are not the same, the amplitude of the resulting wave-form will tend to be *symmetrically* pushed both above and below the *x*-axis. This may result in some strange-looking waveforms but, over time, the average will come out to zero. In effect, even though $g(t)$ is a function of time, it will have the same effect as if it were the simple constant A.

So what about the case when $\eta = \omega$? If we substitute ω for η in Equation 7-4:

$$\int_0^{2\pi k} \sin \omega t \sin \omega t\, dt = \int_0^{2\pi k} \sin^2 \omega t\, dt \neq 0 \qquad \text{Equation 7-5}$$

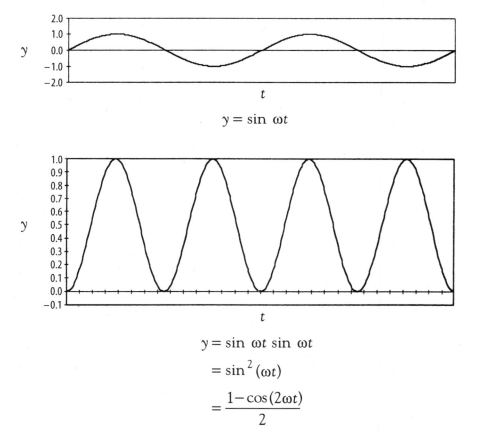

$$y = \sin \omega t$$

$$y = \sin \omega t \sin \omega t$$

$$= \sin^2(\omega t)$$

$$= \frac{1 - \cos(2\omega t)}{2}$$

Figure 7-2: The average of the square of a sine wave is greater than zero.

That is, we get the sum of the *square* of the sine wave. When we square the sine waveform, we get a figure like the one shown in Figure 7-2. Since a negative value times a negative value gives a positive value, the negative portion of the original sine wave is moved vertically above the *x*-axis. The resulting waveform is always positive, so its average value will *not* be zero.

So far the discussion has made use of analytical functions which are useful in developing algorithms and theoretical concepts. As a practical matter, however, in DSP work we are generally more interested in testing a sequence of numbers (the sampled signal) for orthogonality. At this point, we need to take a slight diversion through the subject of continuous functions versus discrete sequences.

Continuous Functions vs. Discrete Sequences

When we look at a function like $y(t) = \sin(2\pi f t)$ we normally think of it as a continuous function of t. If we were to graph the function, we would compute a reasonable number of points and then plot these points. Next, we would draw a continuous and smooth line through all of the points. We would therefore have a continuum of points for t, even though we computed the value of the function at a finite number of discrete points.

In general, we can apply numerical techniques to compute a value for any specific function. For example, even if we cannot analytically solve an integral, we can still compute a specific value for it. From Equation 3-16:

$$\int f(x)dx \approx \sum f(x)\Delta x \qquad \text{Equation 7-6}$$

87

We point this out because it would seem reasonable, when dealing with DSP functions, to adopt the same computational methods. Interestingly enough, we generally do *not*. This fact is not usually emphasized in most texts on DSP, and it can lead to some confusion. While there is not normally a large leap between continuous and discrete functions in mathematics, it often appears that there is some mysterious difference between discrete and continuous functions in DSP. In fact, the discrete and continuous forms of functions used in DSP often *are* different, and therefore have different properties.

Here's why: In Equation 7-6 we can think of both sides of the equation as finding the area under the curve f. Whether or not we find this area by analytically solving the integral, and then evaluating the resulting function, or by numerically evaluating the right-hand side, we expect to get *essentially* the same answer.

Most DSP applications involve an intensive amount of computation. Anything that can be done to save computation effort is important. Furthermore, it turns out that we are often only interested in *relative* values. In most DSP applications the Δx term is really just a scale factor. For these reasons, we often drop the multiplication by Δx. Thus, it is common to see things like:

$$y_c = \int f(x)\,dx \text{ (the continuous form)}$$

and

$$y_d = \sum f(x) \text{ (the discrete form)}$$

Now, these two forms will *not* give us numerically equivalent results. However, surprisingly often, we don't really care. We will demonstrate this concept next as we develop the idea of orthogonality for discrete sequences.

Orthogonality Continued

The discrete form of Equation 7-3 is generally written as:

$$\sum_{n=-\infty}^{\infty} x[n] \sin\left(2\pi f \frac{n}{N}\right) = 0, \text{ if } x[n] \neq \sin\left(\frac{2\pi fn}{N}\right)$$

Equation 7-7

What is the significance of all this? Well, it provides us with a means of testing to see if the sequence $x[n]$ was generated from $\sin(2\pi fn/N)$. This may not seem particularly useful, and in fact, in this form it is *not* particularly useful. This is the case because we need to know the exact phase of $x[n]$ to make Equation 7-7 work. If we could remove this restriction, then Equation 7-7 would have more utility. It would allow us to test to see if the sequence $x[n]$ contained a frequency component at the frequency f. (The importance of this will be made clear in the next chapter.)

We would now like to remove the requirement that $x[n]$ be in phase with the sine function. This is where our next key building block comes into play: *quadrature*.

Quadrature

The term quadrature has a number of meanings. For our purposes the term is used to refer to signals that are 90° out of phase with each other. The classic example of a quadrature signal is the complex exponential:

$$e^{j\omega} = \cos\omega + j\sin\omega$$

This suggests that the complex exponential may be useful in our quest to come up with a more usable form of Equation 7-7. If we multiplied the sequence $x[n]$ by the complex exponential instead of just the sine function, then we would have a complex sequence.

Since a complex number has both phase and magnitude, this allows us much more flexibility in dealing with the phase of the sequence $x[n]$.

To illustrate this concept, take a look at Figure 7-3. The first of three possible phase relationships for the sequence $x[n]$ is shown. In this case the sequence $x[n]$ is in phase with the imaginary part of $e^{j\omega}$. Figure 7-3a shows the imaginary part, and Figure 7-3b shows the real part of $e^{j\omega}$. Figure 7-3c is the function for the sequence:

$$x[n] = \sin\left(\frac{\omega n}{N}\right)$$

Equation 7-8

Now comes the interesting part. Multiplying Figure 7-3a by Figure 7-3c point by point and summing yields:

$$\sum x[n]\, \text{Im}\left(e^{j\omega n/N}\right) > 0$$

Equation 7-9

and the real part is:

$$\sum x[n]\, \text{Re}\left(e^{j\omega n/N}\right) = 0$$

Equation 7-10

We can see this by simply looking at the graphs in Figure 7-3d and Figure 7-3e. In Figure 7-3d we see two interesting features. First, the frequency has doubled. This is not particularly relevant to our current argument, but it is a nice check: from any trigonometry book we know that squaring a sine wave should double the frequency. The second, and more relevant, point is that the waveform is offset above the x-axis. This means that the waveform has some average value greater than zero.

In Figure 7-3e we see that the waveform is symmetrical about the x-axis. Thus, the average value is zero for the real product.

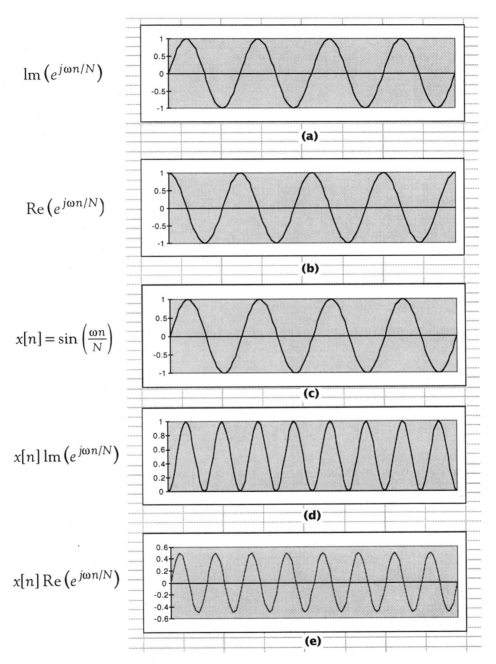

Figure 7-3: Orthogonality: imaginary part in phase.

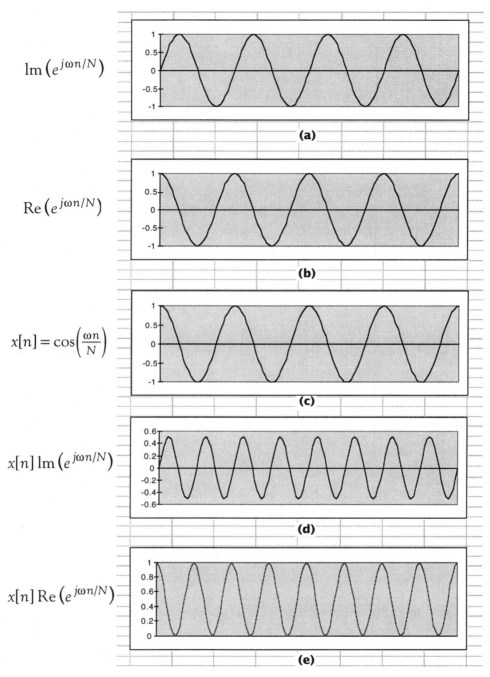

Figure 7-4: Orthogonality: real part in phase.

Figure 7-4 shows the opposite case. In this case, our input function (Figure 7-4c) is:

$$x[n] = \cos\left(\frac{\omega n}{N}\right)$$

Equation 7-11

The sequence $x[n]$ is in phase with the real part of $e^{j\omega}$. In this case:

$$\sum x[n] \, \mathrm{Re}\left(e^{j\omega n/N}\right) > 0$$

Equation 7-12

as shown in Figure 7-4e.

Now, the *really* interesting part of all of this is shown in Figure 7-5. In this case, the sequence $x[n]$ is 45° (or, equivalently, $\pi/4$ radians) out of phase with both the real and imaginary parts of $e^{j\omega}$. At first, this may seem a lost cause. However, in this case, the $x[n]$ lies in the first quadrant. Therefore, a portion of the signal will be mapped into the real sum of the products and a portion of the signal will be mapped into the imaginary portions of the sum of the products, as shown in Figure 7-5d and Figure 7-5e.

Figure 7-5e clearly shows this. Each has a value less than the equivalent case when the input signal was in phase with the real or imaginary part. On the other hand, the value is clearly greater than zero.

We are really only interested in the magnitude of the signal, however, so we can take the absolute value of the sum:

$$\left| \sum x[n] \, e^{j\omega n/N} \right| > 0$$

Equation 7-12

The key point here is that the magnitude of the complex sum is the same regardless of the phase of $x[n]$ with respect to $e^{j\omega}$.

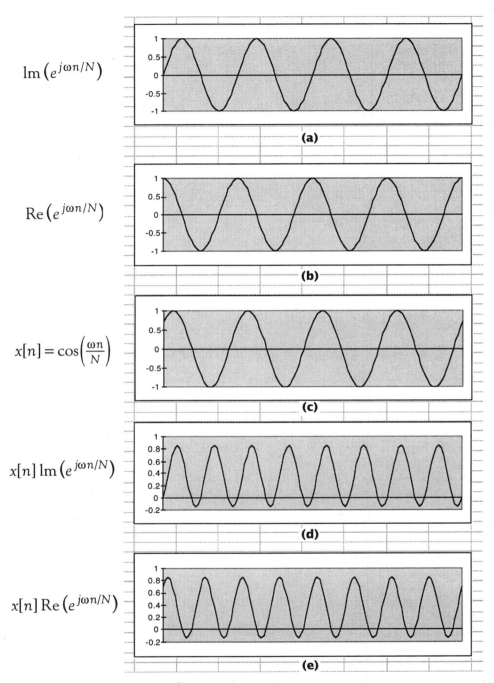

Figure 7-5: Orthogonality: quadrature.

To summarize what we have just done, if we multiply a sinusoidal signal by another sinusoidal signal of the same *frequency* and *phase*, we can tell if two frequencies are the same. We can tell this because the average value of the product will be greater than zero. (OK, we could tell that just by looking at the two signals, too.)

We can eliminate the problem with the phase by multiplying the input function by the complex exponential. When we do this, it does not matter what the phase of the input signal is: part of the signal will map into the real product, and part of the signal will map into the imaginary product. By taking the absolute value of the complex product, we get the same value as if the signal were in phase with one of the real or imaginary parts.

Chapter Summary

Orthogonality, as it applies to most DSP work, simply means that multiplying two orthogonal sequences together and taking the sum of the resulting sequence yields a result that is zero. If the multiplication and addition is done numerically, the result may not be *exactly* zero, but it will be close to zero with respect to the amplitude of the functions.

Orthogonality suggests some useful applications, and these are presented in later chapters. By itself, however, the orthogonality of real functions is of limited value because of an implicit assumption that the two functions (or sequences) are in phase with respect to each other. By using sequences of complex numbers, however, we can bypass the requirement that the functions be in phase. The use of complex numbers in this way is often referred to as quadrature.

This chapter has been one of the more esoteric ones. If you understand the material presented here, then you are definitely ready to move on to the rest of the book. If it does not makes sense to you, you have a couple of options. First, this type of calculation is easily handled by spreadsheets. You can take a look at Chapter 11 for a discussion of using spreadsheets for DSP calculations. The next chapter provides a spreadsheet example based on the material presented here. Setting up a spreadsheet and working through the example will often make these concepts clear. Something to keep in mind is that this material is here to build a base for the subjects in the following chapters. It might be useful to read ahead, and then come back to this section to provide some perspective on orthogonality and quadrature.

8

Transforms

Introduction

In this section we will look at what transforms are and why they are of interest. We will then use the previous discussion on orthogonality and quadrature to develop some useful transforms and their applications. In the next chapter, we will make use of the tools developed in this chapter to design practical digital filters.

Background

In general, a mathematical transform is exactly what the name implies: it transforms an equation, expression, or value into another equation, expression, or value. One of the simplest transforms is the logarithmic operation. Let's say, for example, that we want to multiply 100 by 1,000. Obviously the answer is 100,000. But how do we arrive at this? There are two approaches. First, we could have multiplied the 100 by 1000. Or we could have used the logarithmic approach:

$$100 \times 1000 = 10^2 \times 10^3 = 10^5$$

The advantage of using the logarithmic approach is, of course, that we only need to add the logarithms (2 + 3) to get the answer. No multiplication is required.

What we have done is use logarithmic operations to *transform* the numbers 100 and 1000 into exponential expressions. In this form we know that addition of the exponents is the same as multiplying the original numbers. This is typically why we perform transforms: the transformed values are, in one way or another, easier to work with.

Another common transform is the simple frequency-to-period relationship:

$$f = {}^1/_P$$

This states that if we know the fundamental period of a signal, we can compute its fundamental frequency—a fact often used in electronics to convert between frequency and wavelength:

$$L = P\lambda$$

where L is the wavelength and λ is the speed of light.

The frequency of a radio wave and its wavelength represent the same thing, of course. But for some things, such as antenna design, it is much easier to work with the wavelength. For others, such as oscillator design, it is simpler to work with the frequency. We commonly transform from the frequency to the wavelength, and the wavelength to the frequency, as the situation dictates.

This leads us to one of the most common activities in DSP: transforming signals. Let's start by looking at a simple example.

Figure 8-1a shows a simple oscillator. If we look at the output of the oscillator as a function of *time*, we would get the waveform shown in Figure 8-1b. If we look at the output as a function of *frequency*, we would get the result shown in Figure 8-1c. Notice that in Figure 8-1c we have shown both the positive frequency f and the negative frequency $-f$.

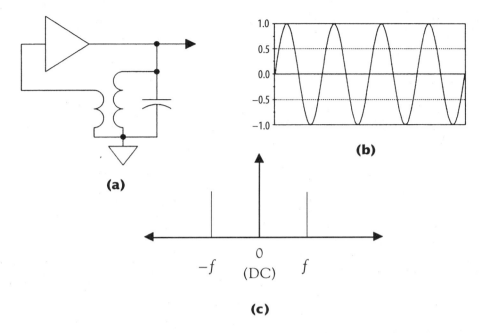

Figure 8-1: Spectrum analysis example.

In most electronics applications, we don't normally show the negative frequency spectrum. The reason for this is that, for any real-valued signal, the spectrum will be symmetrical about the origin. Notice that in Figure 8-1c we can determine both the frequency and the amplitude of the signal. We get the frequency from the distance from the origin and, of course, we get the amplitude from the position on the y-axis.

In this simple case, it was easy to move from the *time domain* (Figure 8-1b) of a signal to the *frequency domain* (Figure 8-1c) because we know the simple relationship:

$$f = {}^1/P$$

Now, what if we wanted to look at the spectrum of a more complicated signal—for example, a square wave?

We can do this by inspection from our work on the Fourier series. We know that a square wave is composed of a sine wave at the fundamental frequency, and a series of sine waves at harmonic frequencies. With this information, we can take a signal like the one in Figure 8-2a and find its spectrum. The spectrum is shown in Figure 8-2b.

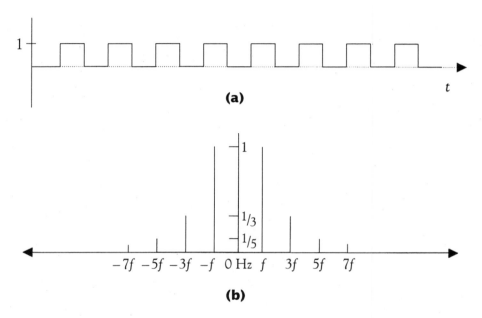

Figure 8-2: Transform of a square wave.

This process of converting from the time domain to the frequency domain is called a *transform*. In this case, we have performed the transform heuristically, using the knowledge we have already developed of the square wave. There are lots of applications for transforms. Often, it is impossible to tell what frequency components are present by simply looking at a the time domain representation of a signal. If we can see the signal's spectrum, however, these frequency components become obvious. This has direct application in seismology, radar and sonar, speech analysis, vibration testing, and many other fields.

With all of these applications, it is only logical to come up with some general-purpose method for transforming a signal from the time domain to the frequency domain (or vice versa).

Fortunately, it turns out that there is a relatively simple procedure for doing this. As you have probably already guessed, it makes use of the techniques from the last chapter: quadrature and orthogonality. Before we move on, however, we need to take a detour through another interesting tool: the z-transform.

The z–Transform

In Chapter 3 we reviewed the Taylor series for describing a function. In that discussion, we pointed out that virtually any function can be expressed as a polynomial series. The z-transform is a logical extension of this concept.

We will start by looking at the variable z, and the associated concept of the z-plane. Next, we will give the definition of the z-transform. We will then take a look at the z-transform in a more intuitive way. Finally, we will use it to derive another important (and simpler) transform: the discrete Fourier transform (DFT).

The variable z is a complex quantity. As we saw in Chapter 3, there are a number of ways of expressing a complex number. While all of the methods are interchangeable, some work better in certain situations than others, and the z-transform is no exception. Thus, the variable z is normally defined as:

$$z = re^{j\omega} \qquad \text{Equation 8-1}$$

In words, any point on the z-plane can be defined by the angle formed by $e^{j\omega}$, located r units from the origin. Or, more succinctly, the point P is a function of the variables r and ω. This concept is shown graphically in Figure 8-3.[1]

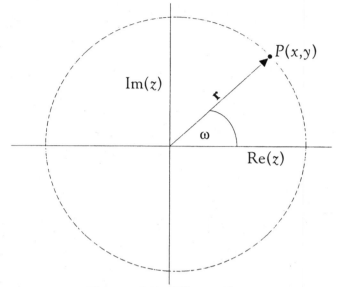

Figure 8-3: The z-plane.

Now, let's look back at the Taylor series:

$$f(x) = \sum_{n=0}^{\infty} a_n x^n$$

This is a real-valued function that expresses the value of $f(x)$ in terms of the coefficients a_n, and the variable x raised to a corresponding power. With only minimal effort, we can generalize this expression to a complex form using Equation 8-1:

---

[1] If this is a little confusing, it might help to compare Figure 8-3 with Figure 3-2. They are really the same thing; only the nomenclature has changed.

$$f(z) = \sum_{-\infty}^{\infty} a_n z^n \qquad\qquad \text{Equation 8-2}$$

where a_n is the input sequence.

Interesting, but what does this have to do with signal processing? Well, as we have seen so far we are normally dealing with signals as sequences of discrete values. It turns out that there are some analytical advantages to using negative values for n, but otherwise it does not make any difference to the overall discussion.

For example, let's say we have an input sequence:

$$a[n] = \{3, 2, 1\}$$

We could express this sequence, using Equation 8-2, as:

$$f[z] = 3z^0 + 2z^{-1} + 1z^{-2} \qquad\qquad \text{Equation 8-3}$$

Now, *why* we would want to do this probably isn't clear, but we will get to this in a minute. In the meantime, let's look at one of the often cited attributes of the z-transform. There is a very interesting property of a series called the shifting property. For example, we could shift the sequence $x[n]$ to the sequence $x[n + 1]$. This would then produce a function:

$$g[z] = 3z^1 + 2z^0 + z^{-1} \qquad\qquad \text{Equation 8-4}$$

Obviously $f[z]$ is not equal to $g[z]$.

For example, if we let $z = 2$, then:

$$f[2] = 3 \times 2^0 + 2 \times 2^{-1} + 1 \times 2^{-2}$$
$$= 4.25$$

and:

$$g[2] = 3 \times 2^1 + 2 \times 2^0 + 1 \times 2^{-1}$$
$$= 8.5 \qquad \text{Equation 8-5}$$

If we look at these two values we might notice that $y[2]$ is equal to half the value of $g[2]$. And, not coincidentally, z^{-1} is also equal to 0.5. In fact, in general:

$$Y[z] = z^{-1}G[z + 1]$$

where the capital letter indicates the z-transform expression of the function. The relationship demonstrated in Equation 8-5 is called the *shifting theorem*.

The shifting theorem is not as mysterious as it might seem at first glance if we remember that multiplying by variables with exponents is accomplished by adding the exponents. Thus, multiplying by z^{-1} is really the same as decrementing the exponent by 1. Indeed, the exponent is often viewed as the index of the sequence —just like a subscript.

The shifting theorem plays an important role in the analytical development of functions using the z-transform. It is also common to see the notation z^{-1} used to indicate a delay. We will revisit the shifting theorem when we look at the expression for the IIR filter.

Now, for a more direct application of the z-transform. As we mentioned earlier, we can think of z as a function of the frequency ω and magnitude r. If we set $r = 1$, then Equation 8-2 reduces to:

$$Y(z) = \sum_{n=-\infty}^{\infty} a_n z^{-n}, \text{ letting } r = 1$$

$$Y[e^{-j\omega}] = \sum_{n=-\infty}^{\infty} a_n e^{-j\omega n/N} \qquad \text{Equation 8-6}$$

The left side of Equation 8-6 is clearly an *exponential* function of the frequency ω. This has two important implications. First, a graph of Y as a function is nearly impossible: it would mean graphing a complex result for a complex variable, requiring a four-dimensional graph. A second consideration is that, effectively, the expression $Y[e^{-j\omega}]$ maps to the unit circle on the z-plane. For example, if we have $\omega = 0$:

$$Y[e^{-j\omega}] = Y[\cos 0 + j \sin 0] = Y[1,0]$$

or if $\omega = \pi/4$, then

$$Y[e^{-j\omega}] = Y\left[\cos \frac{\pi}{4} - j \sin \frac{\pi}{4}\right] = Y\left[\frac{\sqrt{2}}{2}, \frac{\sqrt{2}}{2}\right]$$

In our discussion of orthogonality, we pointed out that the function Y, because it is complex, has information about both the phase and magnitude of the spectrum in the signal. Sometimes we care about the phase, but often we do not. If we do not care about the phase, then we get the amplitude by taking the absolute value of Y.

We can make a further simplification to Equation 8-6. It is acceptable to drop the $e^{-j\omega}$ term and express Y simply as a function of ω. Therefore, we generally express Equation 8-6 as:

$$Y(\omega) = \sum_{n=-\infty}^{\infty} x[n]e^{-j\omega n/N} \qquad \text{Equation 8-7}$$

Believe it or not, we are actually getting somewhere. Notice that the right side of Equation 8-7 is familiar from our discussion of orthogonality. With this revelation we can translate the action of Equation 8-7 into words:

*Let's assume we have an input signal sequence {x[n]}.
We can determine if the signal has a frequency component
at the frequency ω by evaluating the sum in Equation 8-7.
If we do this for values of ω ranging from −π to π we will
get the complete spectrum of the signal.*

Equation 8-7, when evaluated at the discrete points $\omega_k = 2\pi k/N$, $k = 0, 1... N-1$, is commonly called the *discrete Fourier transform* (*DFT*). It is one of the most common computations performed in signal processing. As we noted above, it allows us to transform a function of time into a function of frequency. Or, equivalently, it means we can see the spectrum of an input signal by running it through the DFT.

Application of the DFT

We will pull this all together with an example. First, we will generate a signal. Since we are generating the signal we will know its spectrum; it's always nice to know the correct answer before setting out to solve a problem. Next, we will use the DFT to compute the spectrum, and then see if it gives the answer we expect.

For this example, we will set everything up using a spreadsheet. We could use the accompanying *DSP Calculator* software, but the spreadsheet gives us more insight into what is happening. Table 8-1 shows how we generate the signal. It is composed by adding together two separate signals:

$$f_n = \sin\left[\frac{2\pi hn}{N}\right], \ h = 2$$

and

$$g_n = (0.5)\sin\left[\frac{2\pi hn}{N} + \frac{\pi}{4}\right], \ h = 4$$

where h is used to denote the frequency in cycles per unit time. Notice that the first component (f) and the second component (g) are out of phase with each other by 90° ($\pi/4$). This will help illustrate why we need to use complex numbers in the computation.

The resulting waveform is shown in Figure 8-4. In Figure 8-5 we can see the spectrum for the signal. We can, of course, draw the spectrum by simple inspection of the two components. But let's see if the DFT can give us the same information via computation.

Table 8-1: Signal generation.

n	f = sin(2π(2)n/N)	g = sin(2π(4)n/N+π/4)/2	f+g
0	0.000	0.354	0.354
1	0.383	0.500	0.883
2	0.707	0.354	1.061
3	0.924	0.000	0.924
4	1.000	-0.354	0.646
5	0.924	-0.500	0.424
6	0.707	-0.354	0.354
7	0.383	0.000	0.383
8	0.000	0.354	0.354
9	-0.383	0.500	0.117
10	-0.707	0.354	-0.354
11	-0.924	0.000	-0.924
12	-1.000	-0.354	-1.354
13	-0.924	-0.500	-1.424
14	-0.707	-0.354	-1.061
15	-0.383	0.000	-0.383
16	0.000	0.354	0.354
17	0.383	0.500	0.883
18	0.707	0.354	1.061
19	0.924	0.000	0.924
20	1.000	-0.354	0.646
21	0.924	-0.500	0.424
22	0.707	-0.354	0.354
23	0.383	0.000	0.383
24	0.000	0.354	0.354
25	-0.383	0.500	0.117
26	-0.707	0.354	-0.354
27	-0.924	0.000	-0.924
28	-1.000	-0.354	-1.354
29	-0.924	-0.500	-1.424
30	-0.707	-0.354	-1.061
31	-0.383	0.000	-0.383
32	0.000	0.354	0.354

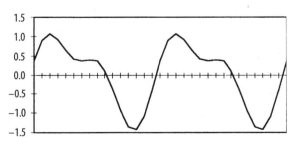

Figure 8-4: Composite waveform.

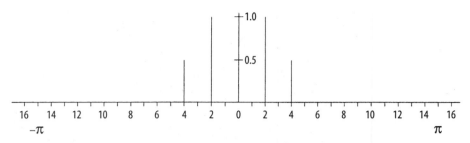

Figure 8-5: Spectrum for the signal in Figure 8-4.

In Table 8-2 we have set up the DFT with a frequency of zero. In other words, we are going to see if there is any DC component. As you can see, the real part of the sum is small and the imaginary part of the sum is zero, so of course the absolute value is small. We can repeat this for any frequency other than $f = 2$ or $f = 4$ and we will get a similar result. So let's look at these last two cases.

Tables 8-2, 8-3, and 8-4 are set up to show the index n in the first column. The second column is the signal $f + g$. The third column is $\mathrm{Re}(e^{-j\omega n/N})$, and the fourth column is $\mathrm{Im}(e^{-j\omega n/N})$. The fifth column is $\mathrm{Re}(f_n e^{-j\omega n/N})$. The sixth column is, naturally, $\mathrm{Im}(f_n e^{-j\omega n/N})$.

For Y[2] we would expect to get a large value, since one component of the signal was generated at this frequency. Since the signal was generated with the sine function, we would expect the value to be imaginary. This is exactly what we see in Table 8-3. The value we get is *not* 1, but by convention, when we plot the spectrum we normalize the largest value to 1.

The actual value in Table 8-3 is 16.0. This is a dimensionless number, not really corresponding to any physical value. If we had used a larger number of samples, the number would have been larger. Correspondingly, a smaller number of samples would have given us a smaller value. By normalizing the value, we account for this variation in the signal length. With this caveat in mind, we can think of the normalized value as the amplitude of the signal.

Table 8-2: DFT with frequency = 0.

n	f+g	cos(2π(0)n/N)	sin(2π(0)n/N	Real Part	Imag. Part
0	0.354	1.000	0.000	0.354	0.000
1	0.883	1.000	0.000	0.883	0.000
2	1.061	1.000	0.000	1.061	0.000
3	0.924	1.000	0.000	0.924	0.000
4	0.646	1.000	0.000	0.646	0.000
5	0.424	1.000	0.000	0.424	0.000
6	0.354	1.000	0.000	0.354	0.000
7	0.383	1.000	0.000	0.383	0.000
8	0.354	1.000	0.000	0.354	0.000
9	0.117	1.000	0.000	0.117	0.000
10	-0.354	1.000	0.000	-0.354	0.000
11	-0.924	1.000	0.000	-0.924	0.000
12	-1.354	1.000	0.000	-1.354	0.000
13	-1.424	1.000	0.000	-1.424	0.000
14	-1.061	1.000	0.000	-1.061	0.000
15	-0.383	1.000	0.000	-0.383	0.000
16	0.354	1.000	0.000	0.354	0.000
17	0.883	1.000	0.000	0.883	0.000
18	1.061	1.000	0.000	1.061	0.000
19	0.924	1.000	0.000	0.924	0.000
20	0.646	1.000	0.000	0.646	0.000
21	0.424	1.000	0.000	0.424	0.000
22	0.354	1.000	0.000	0.354	0.000
23	0.383	1.000	0.000	0.383	0.000
24	0.354	1.000	0.000	0.354	0.000
25	0.117	1.000	0.000	0.117	0.000
26	-0.354	1.000	0.000	-0.354	0.000
27	-0.924	1.000	0.000	-0.924	0.000
28	-1.354	1.000	0.000	-1.354	0.000
29	-1.424	1.000	0.000	-1.424	0.000
30	-1.061	1.000	0.000	-1.061	0.000
31	-0.383	1.000	0.000	-0.383	0.000
			sum =	0	0
			abs(sum) =	0	

What can we expect for the transform of the second frequency component? Since the first component had a non-normalized value of 16, we would expect the second frequency component to have a value of 8. Further, since the second component was generated with a $\pi/4$ phase shift, we would expect this value to be distributed between the imaginary and the real components.

Table 8-3

n	f+g	cos(2π(2)n/N)	sin(2π(2)n/N)	Real Part	Imag. Part
0	0.354	1.000	0.000	0.354	0.000
1	0.883	0.924	0.383	0.815	0.338
2	1.061	0.707	0.707	0.750	0.750
3	0.924	0.383	0.924	0.354	0.854
4	0.646	0.000	1.000	0.000	0.646
5	0.424	-0.383	0.924	-0.162	0.392
6	0.354	-0.707	0.707	-0.250	0.250
7	0.383	-0.924	0.383	-0.354	0.146
8	0.354	-1.000	0.000	-0.354	0.000
9	0.117	-0.924	-0.383	-0.108	-0.045
10	-0.354	-0.707	-0.707	0.250	0.250
11	-0.924	-0.383	-0.924	0.354	0.854
12	-1.354	0.000	-1.000	0.000	1.354
13	-1.424	0.383	-0.924	-0.545	1.315
14	-1.061	0.707	-0.707	-0.750	0.750
15	-0.383	0.924	-0.383	-0.354	0.146
16	0.354	1.000	0.000	0.354	0.000
17	0.883	0.924	0.383	0.815	0.338
18	1.061	0.707	0.707	0.750	0.750
19	0.924	0.383	0.924	0.354	0.854
20	0.646	0.000	1.000	0.000	0.646
21	0.424	-0.383	0.924	-0.162	0.392
22	0.354	-0.707	0.707	-0.250	0.250
23	0.383	-0.924	0.383	-0.354	0.146
24	0.354	-1.000	0.000	-0.354	0.000
25	0.117	-0.924	-0.383	-0.108	-0.045
26	-0.354	-0.707	-0.707	0.250	0.250
27	-0.924	-0.383	-0.924	0.354	0.854
28	-1.354	0.000	-1.000	0.000	1.354
29	-1.424	0.383	-0.924	-0.545	1.315
30	-1.061	0.707	-0.707	-0.750	0.750
31	-0.383	0.924	-0.383	-0.354	0.146
			sum =	0.000	16
			abs(sum) =	16	

In Table 8-4 we evaluate Y[4], and we see that we get exactly what we would expect.

Table 8-4

n	f+g	cos(2π(4)n/N)	sin(2π(4)n/N)	Real Part	Imag. Part
0	0.354	1.000	0.000	0.354	0.000
1	0.883	0.707	0.707	0.624	0.624
2	1.061	0.000	1.000	0.000	1.061
3	0.924	-0.707	0.707	-0.653	0.653
4	0.646	-1.000	0.000	-0.646	0.000
5	0.424	-0.707	-0.707	-0.300	-0.300
6	0.354	0.000	-1.000	0.000	-0.354
7	0.383	0.707	-0.707	0.271	-0.271
8	0.354	1.000	0.000	0.354	0.000
9	0.117	0.707	0.707	0.083	0.083
10	-0.354	0.000	1.000	0.000	-0.354
11	-0.924	-0.707	0.707	0.653	-0.653
12	-1.354	-1.000	0.000	1.354	0.000
13	-1.424	-0.707	-0.707	1.007	1.007
14	-1.061	0.000	-1.000	0.000	1.061
15	-0.383	0.707	-0.707	-0.271	0.271
16	0.354	1.000	0.000	0.354	0.000
17	0.883	0.707	0.707	0.624	0.624
18	1.061	0.000	1.000	0.000	1.061
19	0.924	-0.707	0.707	-0.653	0.653
20	0.646	-1.000	0.000	-0.646	0.000
21	0.424	-0.707	-0.707	-0.300	-0.300
22	0.354	0.000	-1.000	0.000	-0.354
23	0.383	0.707	-0.707	0.271	-0.271
24	0.354	1.000	0.000	0.354	0.000
25	0.117	0.707	0.707	0.083	0.083
26	-0.354	0.000	1.000	0.000	-0.354
27	-0.924	-0.707	0.707	0.653	-0.653
28	-1.354	-1.000	0.000	1.354	0.000
29	-1.424	-0.707	-0.707	1.007	1.007
30	-1.061	0.000	-1.000	0.000	1.061
31	-0.383	0.707	-0.707	-0.271	0.271
			sum =	5.657	5.657
			abs(sum) =	8	

Hopefully, this discussion has been sufficiently clear to demonstrate the basics. If it seems a little fuzzy, it is probably a good idea to work it through. Using a spreadsheet application, try to reproduce the tables in this section. Try different values. A little bit of this kind of work will usually help bring the concepts into clearer focus.

In later chapters, we will see additional uses for the DFT. But for now, let's just point out some characteristics of the DFT. First, the DFT works in both directions: if we feed the spectrum of a signal into the DFT, we will get the time domain representation of the signal out. We may have to add a scaling factor (since we normalized the DFT). Sometimes the DFT with this normalizing factor is called the *inverse discrete Fourier transform (IDFT)*. (Remember that this inversion applies only to the DFT. It is *not* true for the more general z-transform.)

Next, we'll look at two other transforms: the Fourier transform and the Laplace transform. Both are covered here briefly. We are discussing them primarily to make some important comparisons to the DFT and their general relationship to signal processing.

The Fourier Transform

Considering that we just discussed the discrete Fourier transform, we might gather that the Fourier transform is simply the continuous case of the DFT. One of the confusing things in the literature of DSP is that, in fact, the DFT is *not* simply the numerical approximation of the Fourier transform obtained by using discrete mathematics. This goes back to our previous discussion about continuous versus discrete functions in DSP.

This is why we approached the DFT via the z-transform. It really is a special case of the z-transform, and therefore the derivation is more direct. In the DFT, as in the z-transform (or any power series representation), we are working with discrete values of the function. When we move to the continuous case of the Fourier transform, we are actually working with the integral of the function. Geometrically, this can be thought of as follows: The discrete form uses *points on the curve* of a function. The continuous form makes use of the *area under the curve*. In practice, the distinction is not necessarily critical. But it can lead to some confusion when trying to implement algorithms from the literature, or when studying the derivation of certain algorithms.

The forms of the DFT and the Fourier transform are quite similar. The Fourier transform is defined as:

$$H(\omega) = \int_{-\infty}^{\infty} f(t)\, e^{-j\omega t} dt \qquad \text{Equation 8-8}$$

The Fourier transform operator is often written as F:

$$H(\omega) = F\big(f(t)\big)$$

or, equivalently:

$$x(t) \Leftrightarrow X(\omega)$$

It is a fairly uniform convention in the literature to use lower-case letters for time domain functions and uppercase letters for frequency domain functions. In this book, this convention is followed.

Properties of the Fourier Transform

Table 8-5 presents a table of the common mathematical properties of the Fourier transform. These properties follow in straightforward fashion from Equation 8-8. For example, Property 1 states that:

$$aH(\omega) = a \int_{-\infty}^{\infty} f(t)\, e^{-j\omega t} dt = F\big(af(t)\big)$$

where a is an arbitrary constant.

It is worth noting that, as with the geometric series discussed in Chapter 3, the shifting operation applies to the Fourier transform:

$$x(t-\tau) \Leftrightarrow e^{-j\omega\tau} X(\omega)$$

This property is rarely used with relationship to the Fourier transform. It is pointed out here because of the significance it plays in the relationship to the z-transform discussion presented earlier.

A number of other properties of the Fourier transform are pointed out in Table 8.5. Some of these properties, such as the homogeneity property discussed above, follow fairly naturally. Other properties, such as convolution, have not yet been discussed in a context that makes sense. These properties will be discussed in later chapters.

Table 8-5: Some properties of the Fourier transform.

Property	Time function $f(t)$	Fourier transform $X(\omega)$
1 Homogeneity	$ax(t)$	$aX(\omega)$
2 Additivity	$x(t) + y(t)$	$X(\omega) + Y(\omega)$
3 Linearity	$ax(t) + by(t)$	$aX(\omega) + bY(\omega)$
4 Differentiation	$\frac{d^n}{dt^n}x(t)$	$(j\omega)^n X(\omega)$
5 Integration	$\int_{-\infty}^{t} x(\iota)dt$	$\frac{X(\omega)}{j\omega} + \frac{1}{2}X(0)\delta(f)$
6 Sine Modulation	$x(t)\sin(\omega_0 t)$	$\frac{1}{2}[X(\omega - \omega_0) + X(\omega + \omega_0)]$
7 Cosine Modulation	$x(t)\cos(\omega_0 t)$	$\frac{1}{2}[X(\omega - \omega_0) - X(\omega + \omega_0)]$
8 Time Shifting	$x(t - \tau)$	$e^{-j\omega\tau}X(\omega)$
9 Time Convolution	$\int_{-\infty}^{\infty} h(t - \tau) x(\tau)dt$	$H(\omega)X(\omega)$
10 Multiplication	$x(t)h(t)$	$\int_{-\infty}^{\infty} X(\omega)Y(\omega - \lambda)d\lambda$
11 Time and Frequency Scaling	$x(\frac{t}{a}), a > 0$	$a(Xa\omega)$
12 Duality	$X(t)$	$x(-f)$
13 Conjugation	$x * (t)$	$X * (-f)$

The Laplace Transform

The Laplace transform is a natural extension of the Fourier transform. Typically, the Laplace transform does not play a direct role in DSP applications. However, it is being discussed here for several reasons.

One reason is simply to provide completeness of the discussion of transforms in general. Another is the fact that the Laplace transform is often used in many electronics applications that have analogous DSP operations. For example, analog filters are often evaluated using the Laplace transform.

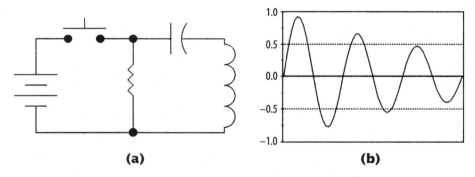

(a)   **(b)**

Figure 8-6: Damped LRC circuit.

Before we look at the Laplace transform, let's consider what inspires us to go beyond the Fourier transform. As noted earlier, the Fourier transform can be used to generate almost any waveform from a series of sinusoidal signals. Some signals, however, are either difficult or mathematically impossible to model efficiently. Consider, for example, the case of an LRC circuit, as shown in Figure 8-6. The general response of this circuit is a second-order differential equation:

$$L\frac{d^2q}{dt^2} + R\frac{dq}{dt} + \frac{q}{C} = v \qquad \text{Equation 8-9}$$

$V(t)$ will have the general solution:

$$Ke^{\tau}e^{j\omega} \qquad \text{Equation 8-10}$$

The circuit response for the underdamped case is also shown in Figure 8-6. Notice that Equation 8-10 simply states what most electrical engineers know intuitively: the response is a damped sine wave. Mathematically, that is a sinusoid multiplied by an exponential function of time. In other words, the output will simply be a "ringing waveform"—a sine wave whose amplitude diminishes exponentially over time.

Solving (or mathematically modeling) something like this with the Fourier transform quickly becomes difficult. The sinusoidal components of the Fourier series are all uniform in amplitude over time. This, naturally, suggests that we expand our definition of the Fourier transform to include an expression something like the one shown in Equation 8-10. This gives us:

$$L\big(x(t)\big) = \int_0^\infty x(t)\, e^{\alpha t} e^{-j\omega t} dt \qquad \text{Equation 8-11}$$

Notice that this is just our definition of the Fourier transform with the addition of the $e^{\alpha t}$ term. In fact, if you set α equal to zero, then Equation 8-11 reduces back to the Fourier transform. Generally, Equation 8-11 is simplified by defining a complex variable $s = \alpha + j\omega$. With this substitution, Equation 8-11 then becomes:

$$L\big(x(t)\big) = X(s) = \int_0^\infty x(t)\, e^{-st} dt \qquad \text{Equation 8-12}$$

This is the classic definition of the Laplace transform. One very interesting aspect of the Laplace transform is that it provides a handy means of solving differential equations, analogous to using logarithms to perform multiplication by adding the exponents.

- First, the individual functions are converted to an expression in the variable s via the Laplace transform.

- Next, the overall system equation is solved *algebraically*.

- Then, the solution is converted back from a variable in s to a variable in t by the inverse Laplace transform.

For example, an inductor become sL, and a capacitor becomes $1/_{sC}$. The loop equation for the circuit shown in Figure 8-6 then can be expressed as:

$$sLI(s) + RI(s) + \frac{1}{Cs} I(s) = V(s) \qquad \text{Equation 8-13}$$

Equation 8-13 is mathematically equivalent to Equation 8-9. Notice, however, that Equation 8-13 is an algebraic expression; there are no differential operators required.

As we noted earlier, the Laplace transform is not often a direct player in DSP applications. Therefore, the development here is kept very brief.[2] In future chapters, however, we will occasionally return to the Laplace transform to make some comparisons and analogies, and to remove some points of confusion between the Laplace transform and the z-transform.

---

[2] For an excellent discussion of the practical use of the Laplace transform, see *Network Analysis with Applications*, by William D. Stanley, Reston Publishing, Reston, Virginia, 1985. For a good general discussion of the Laplace transform as it applies to engineering, see *Complex Variables and the Laplace Transform for Engineers*, by Wilbur R. LePage, Dover Publications, Inc., New York, 1961.

Chapter Summary

In this chapter the concept of orthogonality and quadrature have been developed into the discrete Fourier transform (DFT). From there, we moved to the Fourier transform. The Fourier transform was shown to map a function of time into a function of frequency. This is just the mathematical equivalent of a spectrum analyzer. The Fourier transform was then expanded into the Laplace transform.

The last two chapters have of necessity been rather mathematically oriented. It was necessary to first build the tools that we will use in the following chapters. Unfortunately, this rather mathematical orientation sometimes makes it hard to grasp the concepts at an intuitive level. As the remaining subjects that actually make use of this material are introduced, it will become easier to see the relevance. At that point, it may be a good idea to come back and reread these sections.

CHAPTER 9

FIR Filter Design

Introduction

In the previous chapters we developed a number of tools for working with signals. In order to keep the discussion as tight as possible, these tools were generally presented in a context where they could be understood independently. Convolution, for example, was presented as a generalization of the moving average filter. In a similar manner, the DFT was shown to be a tool that mapped a function of time (the signal) to a function of frequency (the signal's spectrum). We also pointed out, though we did not demonstrate it, that the DFT was a reversible function: given a signal's spectrum, we could use the DFT to get the signal.

It is now time to start tying these tools together to develop a more sophisticated methodology for filter design. Actually, we have all the parts, so let's see how we can arrange them to make a practical design.

Normally, we think of a filter as a function of *frequency*. That is, we draw a graph showing what frequencies we want to let through and what frequencies we want to block. Such graphs are shown in Figure 9-1, where we show the three most common types of filters: the low-pass, bandpass, and high-pass filter.

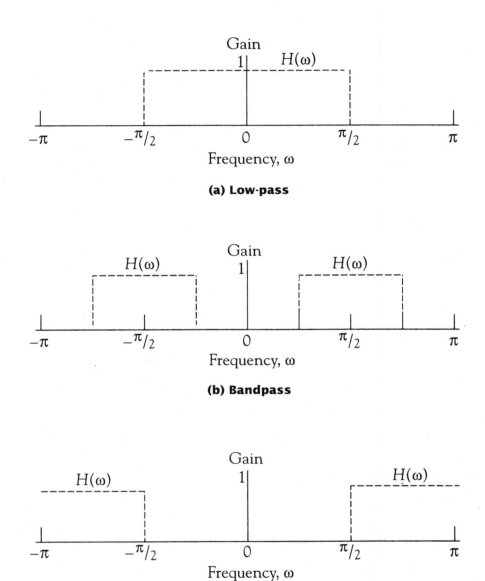

(a) Low-pass

(b) Bandpass

(c) High-pass

Figure 9-1: Three standard filters.

In Chapter 5 we looked at the simple moving average filter. We saw how we could implement it as a convolution of the input signal $x[n]$ with the filter function $h[k]$, where $h[k] = 1/k$. We found $h[k]$ by a purely intuitive process. However, we could also find the function $h[k]$ directly from the DFT.

This provides us with a simple and direct way of generating a filter: we define a filter as a function of frequency $H[\omega]$. We then use the DFT to convert $H[\omega]$ to the sequence $h[k]$. Convolving $h[k]$ with $x[n]$ will then give us our filter output $y[n]$! This is another way of looking at the corollary that convolution in the time domain is equivalent to multiplication in the frequency domain. We will look at a practical example, using the *DSP Calculator* software, shortly. First, however, let's point out that a filter of this type is called a *Finite Impulse Response* filter, or FIR. Let's explore a few of the characteristics of the FIR.

What is an FIR Filter?

The simplest example of a causal FIR filter is our simple moving average filter. As we noted in Chapter 5, the moving average filter can be generated by convolving the input sample $x[n]$ with the transfer function $h[n]$. In the general form, an FIR filter then is:

$$y(n) = \sum_{n=0}^{L} h(m)x(n - m) \qquad \text{Equation 9-1}$$

where L is the length of the filter, and m and n are indexes.

FIR filters get their name from—naturally enough—the way they respond to an *impulse*. For our definition, an impulse is an input of value 1 lasting just long enough to be sampled once and only once. If the response of the filter *must* be finite, then the filter is an FIR. From a practical point of view, a finite response means that, when excited by a unit impulse, the filter's output will return to zero in a reasonable amount of time.

Our simple averaging filters are examples of non-causal FIR filters; given an impulse input, the output will eventually return to zero. As long as the response *must* return to zero for an impulse input, the filter is classified as an FIR. The other major type of filter is the *Infinite Impulse Response* (IIR) filter. As we will see, an IIR filter *may* return to zero for an impulse response, but its architecture does not *require* this to happen.

One helpful way of looking at an FIR filter is shown in Figure 9-2. This type of architectural drawing is generally called a *flow diagram*. As the name implies, a flow diagram sketches the flow of the signal through the system. Notice that the input sequence is shown in what may—intuitively—appear to be the reverse order. In practice, this format is simply showing that f_0 is the first sample of the input sequence. The opposite, but more common, convention is used on the output sequence y.

Several other things in Figure 9-2 deserve comment. The square boxes represent multiplication and the arrows represent delay. Each box is commonly called a *tap*. In this drawing, we have been careful to show two outputs. The output on the bottom of the box is the product of the input sequence and $h(n)$. For the first box, and the

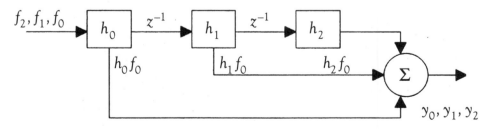

Figure 9-2: Standard architecture for an FIR filter.

first computation cycle, this would be $h_0 f_0$. The output from the right side of the box is just the input delayed by one cycle time. The output of the second box would be $h_1 f_0$ after the second cycle of computation.

The symbol z^{-1} is the standard notation for a unit delay. The circle represents summation, and the output of the summation is the output of our filter.

The simple averaging filter from the last chapter is implemented by setting $h(n) = \frac{1}{3}$ for $n = 0, 1, 2$. Notice that the flow diagram then exactly mimics both the simple averaging routine and the more elaborate convolution sum. It is also worth noting that the flow diagram works equally well for either a software or a hardware implementation. Normally, an FIR filter is implemented in software. However, for systems that require the fastest performance, there is no reason that the multiplication and addition cannot be handled by separate hardware units.

In the real world, when we sit down to design a filter we are usually most concerned with the frequency response. Other considerations are also important, but they are generally second-order concerns. These additional characteristics include such things as the stability of the filter, phase delay, and the cost of implementing the filter. It is worthwhile to look at these second-order concerns before we proceed to a discussion of designing with FIR filters.

Stability of FIR Filters

One of the great advantages of the FIR filter is that it is inherently stable. What this means in practice is that, regardless of what signal we feed into an FIR filter or how long we feed the signal in, when we set the input to zero the output will eventually go to zero.

This conclusion becomes obvious when we think through what the filter is doing. Since it is just multiplying and adding up various parts of the input signal, it follows that the products will eventually all be zero after the last element of the input signal propagates through the filter. This also makes it easy to figure out what the worst-case delay through the filter will be. It is simply the number of taps times the sample rate.

As we will see, this inherent stability is not universal to all digital filters.

Cost of Implementation

The cost of implementation is not just a matter of dollars. The cost is also measured in the resources required and in how long it takes these resources to do the job.

For example, as we mentioned earlier, it is possible to improve the response of an FIR filter by simply increasing the number of taps we use. This has several important consequences, however. First, the more taps we use, the longer it takes to compute the output. For a real-time system, this computation must be completed in less than one sample interval. Further, the more taps we use, the greater the phase delay of the filter. Also of concern is the rounding error. The more computations we make, the more likely round-off errors will increase beyond a reasonable limit.

These factors suggest that we would like to get our output at a minimum cost in terms of the number of computations. The FIR filter is not always the best approach when it is important to minimize computation cycles. On the other hand, the simplicity of designing an FIR filter, combined with its inherent stability, make the FIR filter the preferred choice for many designers.

FIR Filter Design Methodology

As we discussed earlier in the chapter, a variety of filters can be implemented by convolving an input sequence with a transfer sequence. The trick is to come up with a transfer sequence that will produce the desired output from the actual input. While it probably is not obvious, we have already developed the tools we need to do this.

In general, the idea behind FIR filter design is to define the transfer function as a function of *frequency*. This function of frequency, generally named $H(\omega)$, is then transformed into a sequence that is a function of time: $h[n]$. The transformation is accomplished by the inverse discrete Fourier transform (IDFT). A filter is implemented by convolving $h(n)$ with the input sequence $x[n]$. The resulting sequence, $y[n]$, is the output of the filter. This process works for either a real-time process or an off-line processing system.

In practice, the sequence described above will not always produce the desired output $y[n]$. Or, more simply, the filter will not always do what we designed it to do. If this is the case, the function $H[\omega]$ or the sequence $h[n]$ will generally be tweaked to obtain the desired output. This whole design process is shown in Figure 9-3.

Theoretically, any realizable filter can be designed using this simple process. In some cases, however, it will turn out that no amount of tweaking will yield a *practical* design. As we discussed in previous sections, an FIR filter implementation may end up requiring a great number of taps. From a practical point of view, a large number of taps often leads to "mushy" or noisy filter response. When this happens, more sophisticated (that is, more complicated) filters can be tried. These are the subject of later chapters.

The easiest way to understand this design method is with an example.

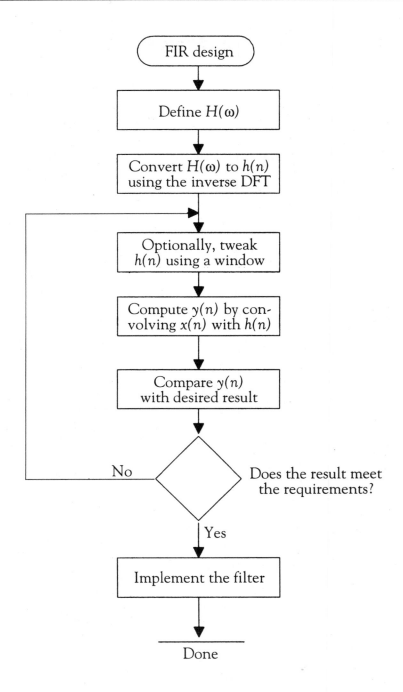

Figure 9-3: Filter design process for an FIR filter.

FIR Design Example

Introduction

The purpose of this section is twofold. First, we will demonstrate the design of a typical DSP application. Second, this application will demonstrate the use of the accompanying *DSP Calculator* software. This example assumes a basic understanding of DSP architecture, convolution, and the discrete Fourier transform. If any of these seem confusing while working through the example, please refer to the appropriate chapters.

For our example, we will design and implement a low-pass filter, requiring the following steps:

- Create a sample waveform with the desired characteristics.

- Look at the spectrum of the sample waveform to ensure that it meets our needs.

- Design the low-pass filter.

- Generate a transfer function to realize the low-pass function.

- Test the design by convolving the transfer function with the sample waveform.

System Description

A block diagram of our system is shown in Figure 9-4. Our system is designed to monitor process signals for an industrial plant. The bandwidth of the signals is 0 Hz to 60 Hz. An anti-aliasing filter is in the front end of the system, and it ensures that any signals will be within this bandwidth.

The signal that we are interested in is a 16-Hz sine wave. Along with this signal is a separate, lower-amplitude, sine wave at 48 Hz.

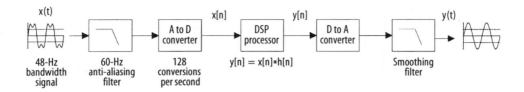

Figure 9-4: Block diagram for low-pass filter example.

Our task is to come up with a digital filter that will keep the 16-Hz signal but eliminate the 48-Hz signal.

Interactive Exercise

Generating a Test Signal

Before we can modify a signal, we must first *have* a signal. Coming up with test signals that have the right characteristics to correctly exercise a DSP system is an important part of the design process. In this case, we can easily generate a test signal using the program *Fourier*. The first thing to do is create a working directory. Use the Windows File Manager to create a directory called *c:\testsig*. Next, open the DSP application group and double click on the icon labeled *Fourier*. Set up the following values in the appropriate boxes:

Frequency: 16

Amplitude: 1

Number of Samples: 128

Then click on the Sin button. You should see a sine wave appear on the screen. Next, set the following values:

Frequency: 48

Amplitude: 0.3333

Number of Samples: 128

Then click the Sin button again. The resulting waveform should look like the one in Figure 9-5. Now save the file to c:\testsig\x.dat. (Use the File / Save command to do this.) Then close the Fourier window; we are done with it for now.

Now we have an input sample with the correct spectral characteristics. The next step is to prove this to be true.

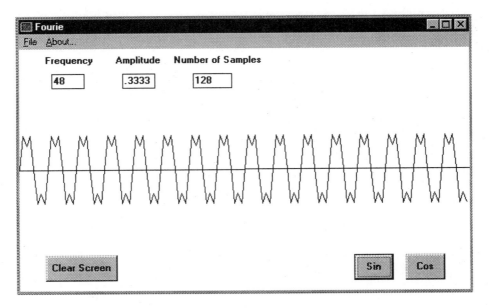

Figure 9-5: Sample waveform for the low-pass filter example.

Looking at the Spectrum

We can look at the spectrum of our signal using the DFT program. Double click on the DFT icon, then load in the file c:\testsig\x.dat. (Use the File / Load Signal menu to do this.) You should see the same wave that was generated in the Fourier program. Now click on the Transform button. Depending upon the speed of your computer, the transformation from the time domain to the frequency domain may take several tens of seconds.

The result should look like Figure 9-6. The first thing to note is that the *x*-axis is the frequency axis. For digitally processed signals, the frequency spectrum is *always* $-\pi$ to $+\pi$. This is called the normalized frequency. Any frequency outside the range of $-\pi$ to $+\pi$ will alias to a frequency with this range. The next logical question is, of course, how does this relate to our actual frequencies?

The answer is that π corresponds to the Nyquist frequency, which is one-half of the sample *rate*. In this example, our sample rate can be assumed to be equal to the *number of samples*: 128. Therefore, the value of π corresponds to a value of 64 Hz. Our base signal is 16 Hz, which is one-fourth of 64. And that is exactly where we see the spectral peak for the 16-Hz signal: one-quarter of the way from 0 Hz to π. We also see the 48-Hz spectral peak at three-quarters of the way to π.

As we would expect, the amplitude of the 48-Hz signal is one-third of the base signal's amplitude. The vertical axis does not really conform to any common physical units such as watts or volts. This is due to the way the transform works. However, the height of the spectral line can loosely be thought of as the amplitude of the signal. The vertical axis is usually scaled to conveniently show the *relative* amplitude of the signals present.

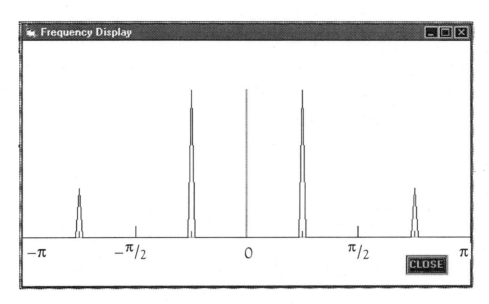

Figure 9-6: Resulting spectrum for the sample signal.

The spectrum is mirrored around the DC (that is, 0 Hz) line. The fact that the negative frequency amplitude components are an exact mirror image tells us that the input signal was either purely real or purely imaginary. Only complex signals can have a positive or negative frequency component that is not symmetrical in amplitude.

This signal meets our spectral requirements for a test signal, so we can now proceed to design and test our filter.

Design the Filter

There are a number of ways to meet our design requirements. Figure 9-7 shows one approach. We have defined a low-pass filter that simply splits the difference between the signal that we want to keep and the signal that we want to reject. So we have set our cutoff frequency at $\pi/2$, or, equivalently, 32 Hz. The filter shape is shown with a dashed line.

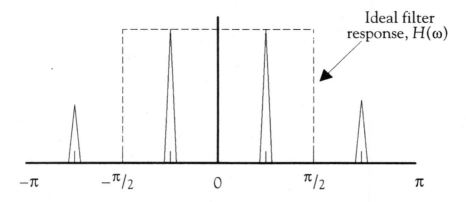

Figure 9-7: Desired filter shape.

Filtering in the frequency domain is a simple operation: we just multiply the frequency components we want to keep by unity. All other frequency components are multiplied by zero. Mathematically, we can define our filter function as:

$$H(\omega) = \begin{cases} 1, -\frac{\pi}{2} < \omega < \frac{\pi}{2} \\ \\ 0, \text{otherwise} \end{cases}$$

Theoretically, we could actually implement our filter this way. We could transform the incoming signal, zero out the 48-Hz spectral line, and then perform an inverse transform to get back to a time domain representation of the signal. Occasionally, filtering is done this way. In general, however, this is extremely inefficient from a computational and implementation point of view. Instead, we make use of the fact that multiplication in the frequency domain corresponds to convolution in the time domain:

$$H(\omega) \bullet G(\omega) \Leftrightarrow h(n) * g(n)$$

Thus, we need to generate a transfer function that, when its impulse response is transformed to the frequency domain, approximates $H(\omega)$. Notice that we said *approximates $H(\omega)$*. Our impulse function will generally not be identical to the ideal case.

We can generate a transfer function by double clicking on the Filter Design icon. Select Filters / Low Pass from the menu. A dialog box with two entries will come up. The first entry is the upper cutoff frequency. We decided in the last section that we wanted a value of $\pi/2$, which is equivalent to 32 Hz in this case. So, enter a value of 1.5708 for the cutoff frequency.

The number of taps determines how closely our filter will approximate the ideal $H(\omega)$. More taps will provide a closer fit. For this example, we will use 15 taps, so enter 15 into the Number of Taps box. Then click on the OK box. The cursor will change to an hourglass, indicating that the transfer function is being computed. (This may take some time if you are working on a slow computer.)

When the computations are done, the hourglass will turn back into the normal cursor and the transfer function will be displayed. Notice that this is the frequency domain representation of the transfer function. Save the transfer function by selecting File / Save As. Save the file to c:\testsig\h.dat. The save operation saves the *time domain* representation of the transfer function. This is also called the impulse response of the transfer function.

Feel free to experiment with the number of taps and with moving the cutoff frequency around, if you like. Close the filter design window when you are done, and we will be ready for the next step.

Convolution of the Signal

We now have our test signal, $x[n]$, and we have just generated a transfer function in the form of $h[n]$. The only thing left to do is to perform the filtering. We will do this by convolving $x[n]$ with $h[n]$.

Double click on the *Convolve* icon. Make sure that the values are set as follows:

Amplitude: 1

Number of Samples: 15

The value of 15 for the *Number of Samples* entry, in this case, corresponds to the number of taps we selected for the filter. Now select *File / Load Coefficients*. Select the file *c:\testsig\h.dat*. The transfer function we generated in the last step will be loaded and displayed. Notice that this is the time domain representation, so it will *not* have the same shape as shown in the filter design program.

Next set the *Number of Samples* to 128. Then select *File / Load Signal* and load the file *c:\testsig\x.dat*. The test signal will be displayed.

Now click on the *Convolve* button. The cursor will turn into an hourglass to indicate that the computations are being performed. A dialog box will appear when the convolution is completed. Click *OK*. The result of the convolution will be displayed. In this case, it is the original 16-Hz sine wave we started with.

We have successfully designed a filter that will allow the 16-Hz signal through, but will block the 48-Hz signal.

For a real application, we would take our transfer function and use it as the h values in the convolution sum:

$$y(n) = \sum_{m=-\infty}^{\infty} x[m]\, h[n-m]$$

Programming the DSP processor to compute this sum would then complete the process.

Windowing

The theoretical definition of the IDFT requires an infinite number of terms to transform $H(\omega)$ into $h(n)$. If we could generate, and make use of, an infinite number of terms from the IDFT we could realize the filter function perfectly. In practice, of course, we must use a finite number of terms for $h[n]$. By truncating the sequence, we are effectively distorting the original function $H[\omega]$. It turns out that we can correct for this distortion of $H[\omega]$ by applying a *compensating* distortion to $h[n]$. This process is sometimes referred to as prewarping the function $h[n]$. Generally, this prewarping process is accomplished by passing $h[n]$ through a *window function*.

There are a number of different windows that can be used. One of the simplest and most common is the *Bartlett window*. The Bartlett window is a simple triangular window. Many other windows exist (rectangular, Hanning, and Hamming, for example), but the idea is the same for all windows: they are "fudge factors" that tweak the coefficients in order to achieve improved performance.

Chapter Summary

In this chapter we defined the general class of filters known as Finite Impulse Response (FIR) filters. These filters are essentially sophisticated versions of the simple moving average filter. An FIR is designed by specifying the transfer function $H(\omega)$. The function $H(\omega)$ is then converted to a sequence using the IDFT. This sequence, $h(n)$, then becomes the coefficients of the filter. The FIR is then realized by convolving the input with $h(n)$.

The FIR filter has a number of significant advantages. It is unconditionally stable, easily designed, and easily implemented. It is possible to design an FIR filter with a linear phase delay.

The one major disadvantage of the FIR is that it can require a large number of computations to implement. A general rule is that an FIR filter should not make use of more than about 30 taps. Beyond this, the response of the filter can get mushy, and the noise caused by truncations can become a problem. However, like all rules of thumb, this one needs to be applied with some caution.

What happens if the number of taps becomes too large? The answer is that we try an infinite impulse response (IIR) filter. This is the subject of the next chapter.

CHAPTER 10

The IIR

Introduction

One of the easiest ways to approach the Infinite Impulse Response (IIR) is to start with the basic equation for the Finite Impulse Response (FIR) and then expand on this base.

If we look back at the basic FIR, we see something like this:

$$y(k) = ax[k] + bx[k-1] + cx[k-2] + \ldots + zx[k-n]$$

Equation 10-1

If we set the coefficients a, b, c ... z equal to $1/(n-1)$, then we have the simple moving average filter. Or we could choose the coefficients according to some function, such as the IDFT of the frequency response of the desired filter, as we did in the last chapter.

Theoretically, *any* filter function can be realized with Equation 10-1. What then motivates us to try something else? The answer is that while any function can be realized with an FIR, there is no guarantee that the function will be realized in an efficient manner. For example, filters with fast roll-offs take a large number of terms to implement. This has two effects: first, the filter algorithm will execute slowly and, second, the delay through the filter may be unacceptably long.

One way to improve the performance of the filter is to make use of the signal values that *have already been processed*. That is, we can make use of previous values of y. For example:

$$y[k] = c[0]x[k] + c[1]x[k-1] + \ldots + c[N]x[k-N]$$
$$+ d[1]y[k-1] + d[2]x[k-2] + \ldots + d[M]y[k-M]$$

Equation 10-2

Notice that we have two sets of coefficients in this form of filter function. One set is called the c coefficients and the other is the d coefficients. If we set the d coefficients equal to zero, then we have our basic FIR filter.

Equation 10-2 is often expressed more compactly as:

$$y[k] = \sum_{n=0}^{N} c[n]x[k-n] + \sum_{m=0}^{M} d[m]y[k-m]$$

Equation 10-3

As a side note, the FIR filter is sometimes called a nonrecursive filter, since it does not make use of the previously processed signal. As one might expect, the IIR is sometimes called a recursive filter since it does make use of previously processed values.

The Infinite Impulse Response (IIR) filter is a little hard to get a handle on in a purely intuitive way. Unlike the FIR, which could be thought of as a modified moving average, the IIR has no convenient intuitive analog. As with the FIR, one of the major attributes of an IIR filter that we are interested in is the frequency response of the filter. In the case of the FIR, we simply took the DFT of the function we were convolving to get the frequency response. Unfortunately, this will not work on Equation 10-2 because Equation 10-2 has both the input and output terms on both sides of the equation. We need a more sophisticated tool than the DFT to handle the situation.

The answer is to make use of the z-transform. This will provide us with all the information we need. If we take the z-transform of each side of Equation 10-2 and rearrange the terms we get:

$$Y(z) - d(1)z^{-1}Y(z) - \ldots - d(M)z^{-M}Y(z)$$
$$= c(0)X(z) + c(1)z^{-1}X(z) + \ldots + c(N)z^{-N}X(z)$$

Equation 10-4

where $Y(z)$ is the transform of the output and $X(z)$ is the transform of the input.

We can define the transfer function as the output of the filter over the input of the filter:

$$H(z) = \frac{Y(z)}{X(z)}$$

Equation 10-5

Now, if rearrange Equation 10-4 into the form of Equation 10-5, we get:

$$H(z) = \frac{c(0) + c(1)z^{-1} + \ldots c(N)z^{-N}}{1 - d(1)z^{-1} - \ldots - d(M)^{-M}}$$

Equation 10-6

This is important because it shows us that the transfer function is the ratio of two polynomials in z. This means that $H[z]$ can vary quickly; the denominator can be used to drive the overall response. This rapid change in $H[z]$ is another way of saying that the filter can have very sharp transition regions, and this can be achieved with far fewer terms than would be required with an FIR filter.

The next step is to rearrange Equation 10-6 into the form of summations, where M is the number of samples we're transforming:

$$H[z] = \frac{\sum\limits_{n=0}^{N} c_n z^{-n}}{1 - \sum\limits_{m=1}^{M} d_m z^{-m}} \qquad \text{Equation 10-7}$$

Now, just as we did with the IDFT, we can find the frequency response by letting $r = 1$ in the definition of $z = re^{j\omega}$. This gives us:

$$H[e^{-j\omega}] = \frac{\sum\limits_{n=0}^{N} c_n e^{-j\omega\frac{n}{N}}}{1 - \sum\limits_{m=1}^{M} d_m e^{-j\omega\frac{m}{M}}} \qquad \text{Equation 10-8}$$

And, just as with the z-transform, this gives us the frequency response as a complex function. It is, in fact, the value above the unit circle in the z-plane. A common practice is to take the absolute value of both sides of Equation 10-9. For simplicity, the resultant function is usually expressed as a simple function of ω:

$$H[\omega] = \left| \frac{\sum\limits_{n=0}^{N} c_n e^{-j\omega\frac{n}{N}}}{1 - \sum\limits_{m=1}^{M} d_m e^{-j\omega\frac{m}{M}}} \right| \qquad \text{Equation 10-9}$$

Or, in other words, we can find out the frequency response of an IIR filter from its coefficients. Since the IIR filter is a ratio of polynomials, the process is more involved than is the case for the FIR filter.

So far, the discussion has followed a more or less standard text-book development. That is, the discussion assumes that we know the coefficients, and that we want to find out what filter response they will give us. The problem with this is that in the real world we generally know what frequency response we *want*. The problem is to come up with the *coefficients*.

The news on obtaining the coefficients for an IIR is mixed. The bad news is that there is no simple and practical way of analytically deriving the coefficients if we are given the desired transfer function. The good news is that there are numerous software tools that make the design of IIR filters relatively straightforward.

Conceptually, an IIR can be designed by starting off with a conventional analog filter. Normally, the filter is expressed in the Laplace form. The Laplacian of the filter is then mapped from the s-plane onto the z-plane. The coefficients of the z-plane representation are then found. This is the approach generally taught in an academic course on DSP filter design. In practice, the process is quite tedious, and not often performed by working engineers.

For practical IIR design, it is generally a good idea to use one of the better filter design software packages. There are a number of reasons for this, mostly centering around the touchy behavior of the IIR. In the case of the FIR filter, we did not have to worry about filter stability, nor did we have to worry a great deal about the phase of the filter. This is not true with the IIR. It is quite possible to design an IIR that has the desired frequency response but is unusable because of stability. It is important to note that even if an IIR is technically stable, it may still exhibit an unacceptable amount of ringing or phase distortion.

With all of these caveats noted, we will now proceed to design an IIR filter. We will design it to meet the same basic requirements as the FIR filter example from the last chapter: a low-pass filter that will pass a frequency at a digital frequency of $\pi/4$, and eliminate a signal at $3\pi/4$. The reason that we are going ahead with the design of the IIR using a somewhat analytical approach needs to be addressed. While we highly recommend the use of professional-quality filter design software for developing digital filters (especially IIRs), designing an IIR from basic principles can illustrate a number of interesting and useful concepts.

Before we proceed, we should discuss the design approach we will be using. This will give us a chance to also look at some key concepts related to the z-transform. Our approach will be to place poles and zeroes appropriately around the z-plane. From the pole/zero graph we will then generate the z-transform in factored form. Next, we will evaluate the partial fraction into a standard polynomial form. From there, we will put the z-transform in the standard form of the definition; then, we can find the coefficients of the IIR by simple inspection.

For this approach to work, we must understand some basic ideas behind the graphical representation of the transfer function in the z-plane. The z-transform of a sequence is complex, as is the function itself. Thus, a graphical representation requires four dimensions. In practice, however, we can obtain a useful graphical image if we look at the absolute value of the transfer function. The absolute value corresponds directly to the amplitude of the transfer function response. We can also find the phase by looking at the angular component, but this is of less interest at this point in the design.

We can think of the absolute value of the transfer function as rubber membrane above the z-plane. The poles of the transfer function are created when any of the factors in the denominator go to zero. Anything divided by zero is undefined, but let's think about what happens as the denominator approaches zero. The transfer function is going to approach infinity, or, in other words, the function will "blow up." Graphically, we can think of the poles as raising up the rubber membrane to an infinite height.

The zeroes in the numerator, on the other hand, produce a value of zero for the transfer function. The zeroes will thus "tack down" the rubber membrane that represents the transfer function. As we noted above, the frequency response of the transfer function is just the z-transform evaluated around the unit circle.

One other key piece of information is required before we proceed. Let's think about what happens on the z-plane. Any point on the z-plane is defined by:

$$z = re^{j\omega}$$

where r is the distance from the origin, and ω defines the angle, relative to the positive real axis. The key concept here, however, is that ω is the angular frequency. We can think about the zero frequency (DC) value lying at (1,0) on the z-plane. The positive frequencies increase, in a counterclockwise direction, until we reach the point (–1,0), which corresponds to an angular value of π. The negative frequencies increase from (1,0) in a clockwise direction until we reach (–1,0).

Now, let's think about what happens when we place a pole directly on the unit circle—at an angle of $\pi/4$, for example. Assuming we start at a frequency of DC, as the frequency increases from DC, we will approach the pole. The denominator will approach 0, and the frequency response of the filter will approach infinity. The filter will blow up; in this case, when the input frequency is $\pi/4$ the output of the filter will be undefined. In practical terms, this means that even a very small input signal at $\pi/4$ (including, for example, a small amount of noise) will cause the output of the filter to try to go to an infinite value. Such a filter is unstable, and therefore probably not of much use to us. This will be true, in fact, for almost any case in which a pole lies on or outside the unit circle. Thus, we have a general rule for filter design: *All poles must lie within the unit circle*.

The corollary to this is that, as the poles move closer to the origin, the amplitude response will decrease, and the general stability of the filter will improve. In general, if we want a sharp filter with high gain we will move the poles as close to the unit circle as practical; if we want a smooth and well-behaved filter, we move the poles as close to the origin as we can get. Note that the relative position of the poles to the zeroes will have a strong effect on the shape of the response.

This all makes more sense if we look at an example. Let's recall the parameters from our FIR example. We have a test signal that is composed of:

$$y[n] = \sin\left[2\pi(16)\frac{n}{N}\right] + \frac{1}{3}\sin\left[2\pi(48)\frac{n}{N}\right]$$

where $n = 0 \ldots N-1$, and $N = 128$. Our sample rate was specified as 128 samples/second. Thus, the digital frequency of the fundamental component of the signal is:

$$\frac{16\,\text{Hz}}{64}\pi = \frac{\pi}{4}$$

and the third harmonic's digital frequency is:

$$\frac{48\,\text{Hz}}{64}\pi = \frac{3\pi}{4}$$

Our design specification was to pass the $\pi/4$ component, and block the $3\pi/4$ component. In the case of the FIR filter, we simply split the difference; we designed a filter that would pass frequencies below $\pi/2$, and block frequencies above $\pi/2$.

For our IIR filter, we can be more specific. We can place the zeroes on the z-plane along the $3\pi/4$ radial. The zeroes will cancel the high-frequency components. To keep the low-frequency components, we will place the poles of the filter along the $\pi/4$ radial. For reasons we will discuss shortly, we will also place a pole at the origin.

First, a couple of general design rules:

- We must maintain symmetry about the x-axis. This will give us the same response for positive and negative frequencies. It will also ensure that the coefficients in the z-transform turn out to be real, and thus the coefficients of the filter will also be real. So, wherever we place a pole or zero, we will also place its complex conjugate on the z-plane.

- To ensure that the resultant filter is causal (that is, we can build a version of it that will run in real time) the order of the denominator must be greater than the order of the numerator.

- As we noted above, all poles must be inside the unit circle to ensure stability.[1] Zeroes can be placed anywhere on the z-plane.

---

[1] Strictly speaking, it is possible to have poles outside the unit circle and still have a stable filter. For example, a zero may cancel out an unstable pole. See *Designing Digital Filters* for a discussion of this.

The key question now, of course, is: Where along the radial to place the poles and zeroes? We will start our selection by making some educated guesses. First, for the poles, a reasonable starting point would be 0.5. We want a little sharper filter, however, so we will set $r = 0.6$. From experience with playing around with this kind of design, we can guess that we will need another pole to smooth out the valley caused by the other two poles. We can achieve this by setting a pole at the origin. This also accomplishes the second design requirement above: it ensures that the denominator will have a higher degree than the numerator, therefore ensuring that our filter will be causal. The zeroes are less of an issue. We can place the zeroes directly on the unit circle at the frequency that we want to suppress.

A pole/zero plot is shown in Figure 10-1.

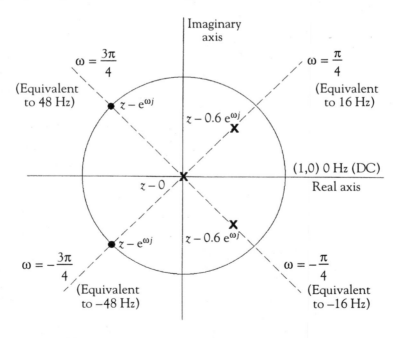

Figure 10-1: Pole/zero plot for the low-pass filter example.

We will develop our filter with the use of one of the standard math packages (see Chapter 11 for a general discussion of design tools). This will give us a chance to explore the use of these tools a little, and it makes our life much simpler. In this case, we will make use of the MathCAD package from MathSoft, Inc. Notice that we are using this package to expedite dealing with the math; we are not using it as a design tool for developing the IIR. The worksheet for the IIR filter is shown in Figure 10-2.

The first thing to note in Figure 10-2 is the initial calculation that we perform at the top of the page. We have chosen a value for N of 40. This needs a little explanation. This value determines the number of points we will *plot* when we look at the frequency response of our filter. That is all it does; it is not related in any way to the sample rate or the coefficients of the filter. Next, we define an index. In this case, n will take on values from 0 to N. We need this index so that we can compute discrete values of the digital frequency. We do this next when we define ω_n, which takes on values from $-\pi$ to π. We then use ω_n to compute the values of z that we will be using in our plot.

The next step is take the poles and zeroes and turn them into the z-transform. We do this by placing the poles in the numerator and zeroes in the denominator. It is fairly straightforward to perform the symbolic computations, but we let the computer do it. First we simplify the numerator, then we simplify the denominator. Notice that we did not try to simplify the entire expression, as this would lead to an unusable and needlessly complex result. As always, we can let the computer do the work, but we cannot let it do the thinking!

At this point we graph $H(z_n)$ to see if it really is close to what we are looking for. Looking at the graph in Figure 10-2, we see that the filter indeed has the frequency response we set out to obtain.

$N := 40$ N is the number of points that we will be using. This number is used for generating the graph of the frequency response. It is not related to number of samples.

$n := 0 .. N$ *n* is the index variable.

$$\omega_n := 2 \cdot \pi \cdot \left(\frac{n - \dfrac{N}{2}}{N} \right)$$ Here, we are computing the indexed digital frequency from $-\pi$ to π.

$z_n := e^{\omega_n \cdot j}$ Next, we compute the value of z at each of the index points.

$$H(z) := \frac{\left(z - e^{\frac{3 \cdot \pi}{4} \cdot j}\right) \cdot \left(z - e^{-\frac{3 \cdot \pi}{4} \cdot j}\right)}{\left(z - 0.6 \cdot e^{\frac{\pi}{4} \cdot j}\right) \cdot \left(z - 0.6 \cdot e^{-\frac{\pi}{4} \cdot j}\right) \cdot (z - 0)}$$ We start the derivation of the transfer function by putting the zeroes over the poles:

$$H(z) := \frac{\left(z^2 + z \cdot \sqrt{2} + 1\right)}{\left(z - 0.6 \cdot e^{\frac{\pi}{4} \cdot j}\right) \cdot \left(z - 0.6 \cdot e^{-\frac{\pi}{4} \cdot j}\right) \cdot (z - 0)}$$ We symbolically evaluate the numerator.

$$H(z) := \frac{\left(z^2 + z \cdot \sqrt{2} + 1\right)}{\left(z^3 - 0.6 \cdot z^2 \cdot \sqrt{2} + .36 \cdot z\right)}$$ Then evaluating the denominator symbolically gives us our transfer function in a usable form.

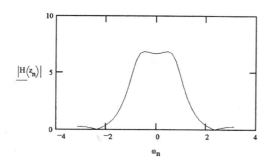

Now we graph the absolute value of the transfer function from $-\pi$ to π to see if it is really what we want. It is. (Remember, since this is a real function, we could have just graphed 0 to π)

Figure 10-2: Electronic worksheet for designing the low-pass filter.

Just for reference, the values of the forty discrete points are as follows:

n	ω_n	z_n	$H(z_n)$	$\lvert H(z_n) \rvert$	$\left\lvert H(z_n) \cdot \dfrac{1}{6.063} \right\rvert$
0	-3.142	-1	-0.265	0.265	0.044
1	-2.985	-0.988 - 0.156j	-0.251 + 0.051j	0.256	0.042
2	-2.827	-0.951 - 0.309j	-0.208 + 0.09j	0.227	0.037
3	-2.67	-0.891 - 0.454j	-0.145 + 0.101j	0.177	0.029
4	-2.513	-0.809 - 0.588j	-0.07 + 0.075j	0.103	0.017
5	-2.356	-0.707 - 0.707j	0	0	0
6	-2.199	-0.588 - 0.809j	0.044 - 0.131j	0.138	0.023
7	-2.042	-0.454 - 0.891j	0.032 - 0.32j	0.322	0.053
8	-1.885	-0.309 - 0.951j	-0.075 - 0.561j	0.566	0.093
9	-1.728	-0.156 - 0.988j	-0.331 - 0.828j	0.892	0.147
10	-1.571	-j	-0.801 - 1.062j	1.331	0.219
11	-1.414	0.156 - 0.988j	-1.555 - 1.137j	1.926	0.318
12	-1.257	0.309 - 0.951j	-2.609 - 0.804j	2.731	0.45
13	-1.1	0.454 - 0.891j	-3.76 + 0.325j	3.774	0.622
14	-0.942	0.588 - 0.809j	-4.287 + 2.533j	4.979	0.821
15	-0.785	0.707 - 0.707j	-3.12 + 5.199j	6.063	1
16	-0.628	0.809 - 0.588j	-0.258 + 6.694j	6.699	1.105
17	-0.471	0.891 - 0.454j	2.832 + 6.26j	6.871	1.133
18	-0.314	0.951 - 0.309j	5.064 + 4.555j	6.811	1.123
19	-0.157	0.988 - 0.156j	6.292 + 2.345j	6.715	1.108
20	0	1	6.675	6.675	1.101
21	0.157	0.988 + 0.156j	6.292 - 2.345j	6.715	1.108
22	0.314	0.951 + 0.309j	5.064 - 4.555j	6.811	1.123
23	0.471	0.891 + 0.454j	2.832 - 6.26j	6.871	1.133
24	0.628	0.809 + 0.588j	-0.258 - 6.694j	6.699	1.105
25	0.785	0.707 + 0.707j	-3.12 - 5.199j	6.063	1
26	0.942	0.588 + 0.809j	-4.287 - 2.533j	4.979	0.821
27	1.1	0.454 + 0.891j	-3.76 - 0.325j	3.774	0.622
28	1.257	0.309 + 0.951j	-2.609 + 0.804j	2.731	0.45
29	1.414	0.156 + 0.988j	-1.555 + 1.137j	1.926	0.318
30	1.571	j	-0.801 + 1.062j	1.331	0.219
31	1.728	-0.156 + 0.988j	-0.331 + 0.828j	0.892	0.147
32	1.885	-0.309 + 0.951j	-0.075 + 0.561j	0.566	0.093
33	2.042	-0.454 + 0.891j	0.032 + 0.32j	0.322	0.053
34	2.199	-0.588 + 0.809j	0.044 + 0.131j	0.138	0.023
35	2.356	-0.707 + 0.707j	0	0	0
36	2.513	-0.809 + 0.588j	-0.07 - 0.075j	0.103	0.017
37	2.67	-0.891 + 0.454j	-0.145 - 0.101j	0.177	0.029
38	2.827	-0.951 + 0.309j	-0.208 - 0.09j	0.227	0.037
39	2.985	-0.988 + 0.156j	-0.251 - 0.051j	0.256	0.042
40	3.142	-1	-0.265	0.265	0.044
(a)	(b)	(c)	(d)	(e)	(f)

Figure 10-3: Table of the computations used in Figure 10-2.

Now that we have the z-transform, it is time to develop the actual filter. The first thing, in this case, is to look at the gain of the filter. We did not specify a gain in our design, but in general we will want a filter that has a gain of 1 at the passband frequencies. Looking at the graph in Figure 10-2, we see that our gain is well above 5. In fact, we can find out exactly what it is if we look at Figure 10-3. In Figure 10-3 we have displayed all of the internal tables that were generated in Figure 10-2. If we look at the frequency response for $-\pi/4$ (i.e, $n = 15$), we see that the frequency response (column e) is 6.063. We want to scale $H(z)$ by the reciprocal of this to give us a gain of 1 at the passband frequency.

The scaled frequency response is shown in column f of Figure 10-3. Now we that we have our scale factor, we can begin to work out the values for the coefficients. The z-transform then is:

$$H(z) = \frac{1}{6.063} \times \frac{z^2 + \sqrt{2}\, z + 1}{z^3 + (0.6)\,\sqrt{2}\, z^2 + 0.36z}$$

which, doing the arithmetic, yields:

$$H(z) = \frac{0.165 z^2 + 0.233\, z + 0.165}{z^3 - 0.849 z^2 + 0.360\, z}$$

Now, we must put this in the form of the definition of the transform:

$$H(z) = \frac{0.165\, z^2 + 0.233\, z + 0.165}{z^3 - 0.849 z^2 + 0.360 z} \times \left(\frac{z^{-3}}{z^{-3}} \right)$$

$$= \frac{0.165\, z^{-1} + 0.233\, z^{-2} + 0.165\, z^{-3}}{1 - 0.849 z^{-1} + 0.360 z^{-2}}$$

Equation 10-10

One of the very nice things about the z-transform is that we can find our coefficients by simple inspection. The numerator gives us the *c* coefficients and the denominator provides the *d* coefficients. The equation for our IIR then is:

$$y[n] = 0.165x[n-1] + 0.233x[n-2] + 0.165x[n-3]$$
$$+ 0.849y[n-1] - 0.360y[n-2]$$

Equation 10-11

Notice that the sign of the coefficients in the denominator is inverted when we put them in the form of the equation.

Several comments are in order on this filter. First, this filter perfectly meets our requirements. That is, it passes the frequencies of $\pi/4$ with a gain of exactly 1, and it blocks signals at $3\pi/4$ completely. In practice, however, this filter is not a particularly good design. It is not very flat in the passband, and the frequency transition is not particularly sharp. We could have done much better by starting with one of the standard analog filters, and mapping the poles and zeroes onto the z-plane. Or, more practically, we could have used a good filter design software package.

Another factor that can cause problems when designing with IIR filters is that the phase of the filter is not linear. Certain frequency components may come out of the filter skewed with respect to other components. All of these factors make it important to carefully evaluate any IIR. The best approach is to use design software to generate plots of the frequency response, phase, group delays, and the pole/zero plots. Remember, the poles of an IIR are the roots of the polynomial in the denominator. The zeroes are the roots of the polynomial in the numerator.

Once the plots for a given IIR look good, it is a good idea to simulate the filter and feed in samples of actual signals. The output of the filter can then be evaluated to see if it will create any problems for the given application.

Chapter Summary

In this chapter we have taken the basic FIR filter and expanded it to the more general IIR filter. The high potential performance of the IIR was noted, but we also pointed out the risks of using the IIR.

A natural question is which to use: the FIR or the IIR? This is a good conversational bomb to drop on a group of DSP experts! Some will argue that, due to the computational efficiency, only IIRs are of any practical use. Others will argue that, due to issues of stability, phase, etc., FIRs are the best choice, with IIRs reserved only for rare cases where the work cannot be handled by an FIR.

In practice, naturally, the decision depends upon the circumstances. FIRs may take 32, 64, or even 128 terms to accomplish a filter requirement. This number of computations may produce an unacceptable loss of precision, especially if the math is done with integers. Or it may simply be too slow. In these cases, it may well be best to go to the IIR. On the other hand, the conceptual, design, and implementation simplicity make the FIR the logical place to start on any design requirement.

CHAPTER 11

Tools for Working with DSP

Introduction

DSP techniques became feasible only when computers were commonly used on a large scale. Therefore, the computer is an extremely important tool for studying, designing, and testing systems based on DSP techniques. For a variety of reasons, this fact was largely ignored by academic courses on DSP in previous years. This is changing, as today even the humblest freshman is likely to have access to computer power that NASA could have only dreamed about 10 or 15 years ago. Academic departments are no longer constrained to dealing with only the analytical approaches to developing a DSP curriculum.

The point of all of this is that the study of DSP can be greatly simplified by use of a good computer and the right tools. The purpose of this chapter is to look at some of the types of tools that are available, and approaches to getting the most out of each tool.

Where practical, names of specific tools are given. Addresses of various sources are provided in the appendix. This listing should not be considered as comprehensive, nor as a recommendation. Software tools come and go. This is particularly true of the tools specifically devoted to DSP. Other, more general, tools will stay around indefinitely—but their applicability to DSP may change.

DSP Learning Software

It is becoming more common for books on DSP to come with some type of programming support. The least useful examples of this are simple listings of C code. This code can be helpful to study and can often be used as a basis for starting a DSP application. On the other hand, code of this type is generally limited in its sophistication. The code is rarely written for a specific compiler or operating system, and therefore is not likely to contain any graphical or display output.

A more useful form of programming support is of the interactive variety. The *DSP Calculator* code included with this book is a good example. *DSP Calculator*'s strong point is that it is intuitive and easy to use. You get immediate visual feedback on the various operations. On the other hand, it is relatively simple, and is not intended for developing DSP applications in a demanding production environment.

Other books also contain executable software. Some of this software provides a great deal of capability, but often at the expense of understandability or ease of use.

DSP learning software is available from many sources other than just books. Commercial training programs, university BBSs, and the BBSs of the DSP processor vendors are often good sources.

Spreadsheets

A number of general-purpose software tools are useful for studying and developing DSP systems. The first of these is the spreadsheet. The examples using spreadsheets in this book are done with Microsoft Excel,™ though most modern spreadsheets will work just as well.

The main use of spreadsheets is in setting up simple series and then manipulating them. The results of these manipulations are then easily graphed. This is how many of the graphs used in the book were generated. It is possible to do very complicated operations with spreadsheets, but it can quickly become more trouble than it is worth. For example, implementing an FIR or an IIR is possible with a spreadsheet, but it requires considerable effort.

On the other hand, a spreadsheet is a great way to see what happens when you multiply two sine waves together on a point-by-point basis, or for studying what happens when sine waves are added together. It's a very convenient way, for example, of getting a feel for the Fourier series.

The big advantage of spreadsheets is that they are available on most PCs, they are simple to use, and they can present the results graphically. For more complicated operations, or when it is useful to actually produce working code, a number of programming languages are available.

Programming Languages

Almost any programming language can be used for studying DSP. For actually producing DSP applications, however, there are three basic choices: assembly language, C, and FORTRAN. It is quite common to use assembly language to produce working applications. Generally, however, coding in assembly is not recommended for studying DSP, or for the early phases of DSP development, due to the tediousness of the programming and the relative difficulty of being able to visualize the algorithm when it is expressed in assembly code. Another difficulty with assembly language programming is that it requires an intimate knowledge of the particular processor being used. Further, depending upon the development environment (see the section later in the chapter), assembly language programming can be difficult to debug and experiment with.

C has become the de facto standard for programming DSP applications and is therefore a natural selection for any type of DSP work. In fact, it is quite common to develop DSP algorithms in a user-friendly environment (such as Borland's C/C++, or Microsoft's Visual C/C++), and then port the application over to the DSP system for the final implementation. Currently, C environments for DSP chips generally follow the ANSI standards. Work is underway, however, that would expand the ANSI C standard to include support of DSP-specific needs. Interestingly, the object-oriented nature of C++ makes it generally less useful for DSP applications. Interest is growing in using C++ for DSP applications, however, and as the language becomes familiar to more programmers, we may see an increase in DSP applications based on C++.

FORTRAN has lost much of its application base to C over the last several years. It is also fairly rare to find a FORTRAN compiler on a PC. For these reasons, the general use of FORTRAN is declining. It is, however, by no means a dead language. Further, a great deal of early program development in DSP was done in FORTRAN. FORTRAN still offers two significant advantages over C: FORTRAN code is simpler to read and write, and FORTRAN has native support for a complex data type.

While these are the three most common working languages for DSP applications, other languages should not be overlooked for studying or developing DSP applications. Probably the best choice, in many regards, is Basic. While Basic is often considered a programmer training language, it has been used in a wide variety of sophisticated applications. Modern Basic languages offer sophisticated error handling, easy graphic implementation, structured program development, and other sophisticated attributes.

For working in the Windows environment, Microsoft's Visual Basic™ is one practical way to write programs without enduring an extremely long learning curve. (The *DSP Calculator* software is written in Visual Basic.) It offers a combination of simple basic programming, an object-oriented interface to Windows, and an affordable price tag.

General Mathematical Tools

A number of good general-purpose mathematical tools are available. Unfortunately, they tend to be fairly expensive. If they are available, however, they are an invaluable aid in studying DSP. Like the other tools discussed so far, these tools automate the frustratingly tedious computations involved in DSP techniques. They have another significant advantage over the other tools discussed so far: they can symbolically evaluate analytical expressions. A good example of this is the IIR example in Chapter 10. Here, both symbolic and arithmetical computations were carried out using MathCAD™ by MathSoft.

Other popular tools include MatLab™ by MathWorks and Mathematica™ by Wolfram Research. Generally, these tools come with optional modules that provide DSP-specific functions. Like any software tool, some effort is required to learn how to use these products effectively. It must also be kept in mind that these tools can unload much of the work involved in developing a DSP application, but they cannot unload the creative part of the design. Nor can they be expected to catch system design errors, or to provide good results if they are fed bad data. These responsibilities still reside with the design engineer. As always, it is necessary to have a firm understanding of the principles and limitations of DSP techniques to obtain meaningful results.

Most of these tools are available in versions that will run on UNIX workstations and on Windows-based PCs. Some versions are available for the Macintosh computer, but these are often inferior to the versions available for other two platforms.

Special-purpose DSP Tools

The quality and availability of these tools vary too much to present a meaningful list here. Often, these packages are customized for specific processors. In fact, not only will these tools give you the desired coefficients, but they will also often give you the assembly language code for implementing the filter. (Since all DSP processors provide special instruction sets designed to implement common DSP operations, this is not as sophisticated a feature as it may seem. On the other hand, every little bit helps when you are on a tight schedule.)

The best way to find out about these packages is through the manufacturer of the DSP processor you plan to use. The manufacturer will generally provide a list of third-party vendors. Another approach is to ask the FAE (field application engineer) that supports the processor in your area for their recommendation on which products to use.

Software/Hardware Development Packages

A wide range of products are available that provide for complete hardware and software development. These products generally come in two flavors: evaluation units and development systems. Evaluation units generally include the DSP processor, program and data memory, and limited analog I/O. They are typically provided by the vendor of the DSP processor. Some of these can be quite reasonably priced (under $100). The price for more sophisticated units can run into thousands of dollars. Evaluation units, as the name implies, are designed to allow you to play with the processor in a real setting to see if it will meet the anticipated requirements. For simpler projects, evaluation units may meet all of your development needs.

There are some things to keep in mind about evaluation units, however. These systems are generally designed to be sold at or below cost to vendor. As such, they generally don't offer a great deal of flexibility. The analog I/O is typically limited, and often some of the processor resources (such as interrupts, internal memory, etc.) are devoted to supporting the evaluation configuration. Evaluation units are generally limited to providing assembly language development tools.

More sophisticated systems are typically available under the heading of development systems. Often, these units are designed as add-on cards that fit into a standard IBM AT clone PC. Typically, compilers, source level debuggers, assemblers, and Host PC interface software are provided. A particularly useful feature of many of these systems is the included library of software routines. Depending upon your application, these libraries may more than pay for the whole development system by shortening design and coding time.

In-circuit Emulators

Another tool available for developing applications is the *in-circuit emulator*. The emulator is an electronic pod with the same pin-out as the target processor. The pod is connected to a PC or UNIX workstation, where the development software resides. Emulators are useful when the actual product hardware already exists. The emulator is plugged into the target hardware instead of the DSP processor, allowing programs to be downloaded to the target system. The pod emulates the processor, but it also provides complete visibility into the operation of the system.

In-circuit emulators are invaluable aids for developing embedded applications. They often greatly shorten the hardware and software debug time. Unfortunately, emulators are also fairly expensive. They start at around $5,000, and go up to $30,000 or more.

World Wide Web

In the last few years the World Wide Web (WWW) has become a major tool for researching a wide variety of topics. The material on the Web that relates to DSP is large, varied, and rather dynamic. As with using the Web to find about any subject, the problem is to find the URL (Universal Resource Locator) that has what you are looking for.

The tools for searching the Web are the *search engines*. Magellan, Yahoo, and InfoSeek are a few examples. Each search engine has its advantages and disadvantages. The trick is to try the various options for each search engine, and if that does not produce what you are looking for, try a different search engine.

Another way to find relevant information is to start with the home pages of the manufacturers of DSP hardware and software. See Appendix B for a list of vendors. Often these pages will have links to other pages that contain useful information.

Chapter Summary

Understanding the concepts behind DSP techniques requires developing practical applications. Even if these applications are simple ones, the benefits to you are twofold. First, the hands-on experience will help solidify the abstract concepts into practical skills. And, second, only by actually developing working applications do you confirm that you do indeed understand the key concepts. The tools discussed in this chapter provide a good starting point to begin your DSP exploration. A little time invested in mastering these tools will return a significant gain in practical knowledge and understanding.

Finally, most of the software and hardware tools discussed here are based on, or are at least available for, the IBM PC clones. It should be remembered, however, that a modern 486DX or Pentium class machine, with its hardware floating point and high-speed peripheral busses, actually makes a decent signal processing machine on its own.

CHAPTER 12

DSP and the Future

One of the continuing themes of this book is that learning DSP techniques can sometimes seem intimidating. This is not so much that the actual techniques are particularly difficult, but the difficulty often lies in the fact that the techniques are, for the most part, purely mathematical. When we set out to learn about analog filters—a simple LC circuit, for example—we have a number of different tools to use. The circuit physically exists, so we can approach it from a physics point of view. We can imagine the currents flowing, the various potentials. We can see these things on an oscilloscope. Only when we need to compute specific values for components must we turn to algebra. Or, if our application requires more detailed analysis, we can bring higher mathematics into play and work with the differential equations that describe the behaviors of the components.

When we turn to DSP techniques it initially seems like we lose this "big picture." Instead, we are faced with obscure integrals, unfamiliar summations, and expressions that are intuitive only to professional mathematicians. As we gain some experience, however, some interesting insights evolve. Among these are the fact that many of the concepts from physics and electrical engineering carry over. Frequency is still frequency (though we have to deal with the *digital frequency*), the same relationships between the frequency domain and time domain still hold, and so forth. In fact, we quickly find that the digital domain is much *simpler* than the analog domain. Differential equations are replaced with simpler difference equations. We are largely removed from the real-world constraints of specific component values, parasitic effects, and the like. This does not mean that no real craftsmanship goes into producing a workable DSP system, but the digital approach is almost always simpler to implement from a conceptual point of view. The hardware may or may not be simpler, but getting a digital system to do what we want is usually simpler than getting an analog system to respond the way we want.

This fact, combined with the fact there are some operations that simply are not practical in the analog domain, ensure that DSP techniques will continue to take over a larger and larger percentage of applications. Even a modern analog-based cellular phone, for example, still contains a significant amount of *digital* circuitry.

The rapid acceptance of DSP techniques is having a snowball effect. Twenty years ago DSP was an esoteric practice largely confined to Ph.D.s. Today, virtually all engineers and a significant percentage of technicians will have at least some exposure to DSP. As the price continues to fall for DSP components, more and more projects are being based around them. In a few short years, proficiency in DSP will not simply be an advantage; it will be a *requirement* of almost any technical position.

Not all of the gains in DSP applications will place a burden on the typical design engineer. A good example is the modern modem chip. These specialized devices are practical marvels of signal processing, and yet designing them into a product requires only a rudimentary understanding of communications theory. The DSP expertise is largely buried in the silicon. For some applications, this trend will continue. Obviously, for these types of products, only the chip designer must be a true expert in DSP. As embedded applications grow, however, the need for a greater understanding of DSP techniques will continue to grow as well.

Finally, the thought I would like to leave with you is that DSP can honestly be fun. The resources of the modern desktop computer make it possible to easily and rapidly experiment with the ideas behind digital signal processing. Once you get past the seemingly formidable nature of the techniques, DSP algorithms are often quite simple and easy to implement.

Remember, even though we live in an analog world, the future is digital!

Fundamentals of Engineering Calculus and Other Math Tools

A.1 Introduction

Even though calculus is taught at the high-school level today, many technically oriented people (engineers among them) still avoid it like the plague. For some reason, many of us—maybe because of past bad experiences with math—feel uncomfortable with calculus. The problem is made worse by the fact that it seems it is always taught by people who know little about engineering. However, whether you like hearing this or not, I have to say it: A good engineer *needs* to be comfortable with calculus. If you don't master this essential tool at some point, you will limit your career.

This appendix is an attempt to present a low-B.S, fog-free review of the essentials of calculus from an engineering systems point of view. I've tried to keep fancy words and theorems to a minimum. I use a familiar engineering system—your automobile—to introduce the concepts of differential and integral calculus.

A.2 Differential Calculus

One of the first helpful things to realize about calculus is that you deal with it every day. If you drive a car, every time you use the gas pedal you are directly applying calculus as you control the car's speed. Assume your car is at a standstill and you tramp down on the gas pedal. Refer to Figure A.1 and let's dig into calculus.

At some time t you will have traveled a distance x from the starting point. We say x is a function of time and write it using symbols like $x(t)$, or $x = f(t)$, where $f(\ )$ means "a function of."

During a small time interval Δt, your car will travel a distance Δx. This incremental change in distance can be expressed in equation form as:

$$\Delta x = x(t + \Delta t) - x(t) \qquad (A.1)$$

That is, the change in distance is equal to the future position $x(t+\Delta t)$ minus the present position $x(t)$.

If both sides of (A.1) are divided by the time interval Δt then

$$\frac{\Delta x}{\Delta t} = \frac{x(t + \Delta t) - x(t)}{\Delta t} \qquad (A.2)$$

If Δx is measured in feet and Δt in seconds, the units of $\Delta x/\Delta t$ are feet per second—otherwise known as velocity. Equation (A.2) is an approximation of the velocity of your car. I'll use the symbol v to indicate velocity.

Let's work with an example. Say the distance your car is away from its starting point is given by the function

$$x(t) = 5t^2 \qquad (A.3)$$

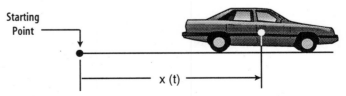

Starting Point

x (t)

Figure A.1. At time *t* the automobile will have travelled a distance *x(t)*.

168

and you want to determine its velocity two seconds after you tramped down on the accelerator (that is at $t = 2$ sec). Using equation (A.2) we write

$$\frac{\Delta x}{\Delta t} = \frac{x(t + \Delta t) - x(t)}{\Delta t} = \frac{5(t + \Delta t)^2 - 5t^2}{\Delta t} \quad (A.4)$$

If we now expand equation (A.4) we get

$$\frac{\Delta x}{\Delta t} = \frac{5(t^2 + 2 \cdot \Delta t \cdot t + \Delta t^2) - 5t^2}{\Delta t}$$

$$= \frac{10 \cdot \Delta t \cdot t + 5\Delta t^2}{\Delta t}$$

or

$$\frac{\Delta x}{\Delta t} = 10t + 5\Delta t \quad (A.5)$$

Now let's apply equation (A.5) to determine the velocity at $t = 2$ using various values for the time increment Δt as shown in Table A.1. You can see from this table that as Δt gets smaller, the velocity approximation $\Delta x / \Delta t$ is approaching 20 ft/sec.

Table A.1.

Δt (seconds)	v(ft/sec) at t = 2 seconds
1.0000	10 x 2 + 5 x (1.0000)² = 25.00000000
0.5000	10 x 2 + 5 x (0.5000)² = 21.25000000
0.2500	10 x 2 + 5 x (0.2500)² = 20.31250000
0.1000	10 x 2 + 5 x (0.1000)² = 20.05000000
0.0100	10 x 2 + 5 x (0.0100)² = 20.00050000
0.0010	10 x 2 + 5 x (0.0010)² = 20.00000500
0.0001	10 x 2 + 5 x (0.0001)² = 20.00000005

From equation (A.5) we can see that

$$\frac{\Delta x}{\Delta t} = 10t \qquad \text{as } \Delta t \to 0 \qquad (A.6)$$

So we go back to equation (A.2) and define the *derivative of a function* $x(t)$ with respect to t as

$$\frac{dx}{dt} = \lim_{\Delta t \to 0} \frac{\Delta x}{\Delta t} = \lim_{\Delta t \to 0} \frac{x(t + \Delta t) - x(t)}{\Delta t}$$

$$(A.7)$$

As you just saw in the example, equation (A.7) can be used to derive the derivative of a function.

Figure A.2 shows a plot of equation A.3.

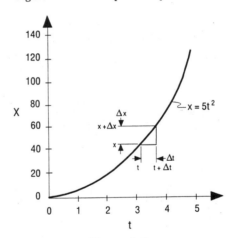

Figure A.2.

You can see that Δx divided by Δt is a slope. As Δt gets smaller and smaller, this slope approaches the derivative $\Delta x / \Delta t$. Therefore, the derivative of function $x(t)$ at the time t is equal to the slope of the function at time t.

Go back now to equation (A.6) and rewrite it as

$$v = \frac{dx}{dt} = 10t \qquad \text{(A.8)}$$

You can see that the velocity v is also a function of time. That is, $v = v(t)$. Since $v(t)$ is just another function of time, we can also take the derivative of this function. That is,

$$\frac{dv}{dt} = \lim_{\Delta t \to 0} \frac{\Delta v}{\Delta t} = \lim_{\Delta t \to 0} \frac{v(t + \Delta t) - v(t)}{\Delta t} \qquad \text{(A.9)}$$

or

$$\frac{dv}{dt} = \lim_{\Delta t \to 0} \frac{10(t + \Delta t) - 10t}{\Delta t} = \frac{10\Delta t}{\Delta t} = 10 \qquad \text{(A.10)}$$

In this case, the derivative of velocity dv/dt (which has the units of feet per second per second, or acceleration) is a constant. Now you know that your car is accelerating at 10 ft/sec², which of course you would feel on your back as you are pressed against the seat.

We twice differentiated the function describing the distance of your car from its starting point. This is called *double differentiation* and can be expressed as

$$\frac{d}{dt}\left(\frac{dx}{dt}\right) = \frac{d^2 x}{dt^2} \qquad \text{(A.11)}$$

This is called taking the second derivative of a function.

We can also write derivatives in shorthand as

$$\frac{dx}{dt} = \dot{x} = v \qquad \text{(A.12)}$$

$$\frac{d^2 x}{dt^2} = \ddot{x} = \dot{v} = a \qquad \text{(A.13)}$$

You can also use the *operator*

$$D = \frac{d}{dt} \qquad \text{(A.14)}$$

to express the derivative. This can be extremely handy. For example, equations (A.12) and (A.13) above can be expressed in operator notation form as follows

$$Dx = \frac{dx}{dt} = \dot{x} = v \qquad \text{(A.15)}$$

$$D^2 x = \frac{d^2 x}{dt^2} = \ddot{x} = \dot{v} = a \qquad \text{(A.16)}$$

Throughout this book block diagrams are used as an aid in building mathematical models. If $x(t)$ is an input or forcing function into the block below and the output is the derivative of the function, then the block must contain the differentiation operator. That is,

is the same as equation (A.15) and

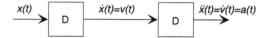

is the same as equation (A.16). The operator D is often called a *differentiator* when used in block diagrams.

A differentiator can also be written for a digital computer. Listing A.1 is a "differentiator" BASIC computer program and Listing A.2 is a spreadsheet version. Carefully review these pro-

Listing A.1.
Digital computer differentiator.

```
5    REM BASIC DIFFERENTIATOR

10   DEF FN X(T) = 5 * T^2

20   INPUT "VALUE OF T, PLEASE"; T

30   INPUT "VALUE OF DELTA T,
     PLEASE"; DELT

40   X1 = FN X(T)

50   X2 = FN X(T + DELT)

60   DELX = X2 - X1

70   XDOT = DELX/DELT

80   PRINT "DERIVATIVE IS"; XDOT

90   END
```

Listing A.2.
Spreadsheet differentiator.

	A	B
1	T	
2	DELT	
3	X(T)	=5*B1^2
4	X(T+DELT)	=5*(B1+B2)^2
5	DELX	=B4-B3
6	XDOT	=B5/B2

grams. I tried to write them as simply as possible to emphasize that differential calculus is in fact simple. Experiment with the programs on your computer. Input various values of Δt while holding t constant to see how it affects the answer that the "digital computer differentiator" provides. Use the programs to experiment with other functions by changing the define function statement in the BASIC program or the statements in cells B3 and B4 in the spreadsheet version.

Even though the programs given in Listings A.1 and A.2 are handy and demonstrate just how easy differential calculus is, I found out early in my career that I saved a lot of time by committing to memory the most frequently used derivatives, provided in Table A.2. You can always look these up in this or other books, but it will take you time and you may not always have your books with you. Every one of these formulas can be derived using equation (A.7), but it's still easier to commit them to memory.

Incidentally, I have been using time t as the *independent variable* and x as the *dependent variable*. That is, $x = f(t)$. I've done this because in many real-life engineering problems, the variables depend on time. However, variables can be a function of another variable that is not time. For example:

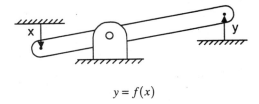

$$y = f(x)$$

Table A.2. Most frequently used derivatives
(where *c* = a constant and
***u* and *v* are functions of *x*).**

$$\frac{dc}{dx} = 0$$

$$\frac{d}{dx}(x) = 1$$

$$\frac{d}{dx}(u+v) = \frac{du}{dx} + \frac{dv}{dx}$$

$$\frac{d}{dx}(cu) = c\frac{du}{dx}$$

$$\frac{dy}{dx} = \frac{dy}{du} \cdot \frac{du}{dx}$$

$$\frac{d}{dx}\log_a u = \frac{1}{u}\log_a e \frac{du}{dx}$$

$$\frac{d}{dx}\ln u = \frac{1}{u}\frac{du}{dx}$$

$$\frac{d}{dx}(u^n) = nu^{n-1}\frac{du}{dx}$$

$$\frac{d}{dx}(uv) = v\frac{du}{dx} + u\frac{dv}{dx}$$

$$\frac{d}{dx}\left(\frac{u}{v}\right) = \frac{v\frac{du}{dx} - u\frac{dv}{dx}}{v^2}$$

$$\frac{d}{dx}\sin u = \cos u\frac{du}{dx}$$

$$\frac{d}{dx}\cos u = -\sin u\frac{du}{dx}$$

$$\frac{d}{dx}\tan u = \sec^2 u\frac{du}{dx}$$

$$\frac{d}{dx}\cot u = -\csc^2 u\frac{du}{dx}$$

$$\frac{d}{dx}\sec u = \sec u \tan u\frac{du}{dx}$$

$$\frac{d}{dx}\csc u = -\csc u \cot u\frac{du}{dx}$$

$$\frac{d}{dx}a^u = a^u\ln a\frac{du}{dx}$$

$$\frac{d}{dx}e^u = e^u\frac{du}{dx}$$

$$\frac{d}{dx}u^v = vu^{v-1}\frac{du}{dx} + u^v\ln u\frac{dv}{dx}$$

or even two variables that are not functions of time

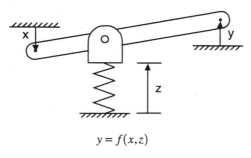

$$y = f(x,z)$$

Nothing changes in the formulas for differentiation except the symbol used for the independent variable. Sometimes you will see a prime symbol (´) used for the shorthand version of differentiation instead of the dot notation. That is,

$$y = f(x)$$

$$\frac{dy}{dx} = y'$$

$$\frac{d^2y}{dx^2} = y''$$

A.3 Integral Calculus

Integral calculus is nothing more than the reverse of differentiation. For example, given that the distance your car is from its starting point is described by $x = 5t^2$, we found that the velocity of the car at any point in time was $v = 10t$ and the acceleration at any point in time was $a = 10$. If integration is the reverse of differentiation, then given the acceleration of the car is 10 ft/sec², we should be able to integrate once and get $v = 10t$ and integrate again and get $x = 5t^2$. Let's look at how we can do this.

We know that the velocity is given by equation (A.8) as

$$v = \frac{dx}{dt} = 10t \qquad \text{(A.8)}$$
$$\text{repeated}$$

The derivative dx / dt is approximately equal to $\Delta x / \Delta t$ when Δt is very small and was given in equation (A.6) as

$$\frac{\Delta x}{\Delta t} = 10t \qquad \text{as } \Delta t \to 0 \qquad \text{(A.6)}$$
$$\text{repeated}$$

We can rewrite (A.6) as

$$\Delta x = v\Delta t \qquad \text{(A.17)}$$

Figure A.3 shows the velocity of the car as a function of time and the graphical representation of equation (A.17).

The increment Δx that your car travels in Δt seconds can be seen to be an incremental area under the $v(t)$ curve. If this represents a small part of x, then to get x at any arbitrary time t, all we should have to do is sum all of these small incremental areas up to that time. This can be expressed using the summation symbol Σ as follows:

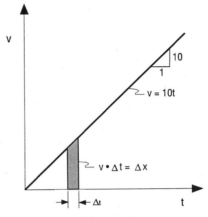

Figure A.3.

$$x \cong \sum_{t=0}^{t=t_1} \Delta x \cong \sum_{t=0}^{t=t_1} v\Delta t \qquad \text{(A.18)}$$

As Δt is made smaller and smaller, the summation symbol (Σ) is replaced by another symbol ($\int$) called the integration symbol and Δx and Δt are replaced with dx and dt. That is,

$$x = \int_{t=0}^{t=t_1} dx = \int_{t=0}^{t=t_1} v\,dt \qquad \text{(A.19)}$$

Since in our example $v = 10t$ we can write

$$x = \int 10t\,dt \qquad \text{(A.20)}$$

We have already seen that one way to evaluate this integral is to find the area under the curve $v(t)$. Another way is to simply find the function whose derivative is $10t$. From our table of derivatives given in Table A.2 we find

$$x = 5t^2 + C \qquad \text{(A.21)}$$

The unknown constant C must be included because when we differentiate equation (A.21) we get

$$\frac{dx}{dt} = 10t \qquad \text{(A.22)}$$

no matter what the value of C is. This constant is called *the constant of integration* or *the integration constant*. It must be evaluated from known conditions. In our car example, we said at $t = 0$ that $x = 0$. Thus,

$$x(t = 0) = 5 \cdot 0^2 + C = C \qquad \text{(A.23)}$$

So C must equal 0 and we arrive at our answer,

$$x = 5t^2 \qquad \text{(A.24)}$$

We can now generalize what we have learned in the following equation

$$\int f(t)dt = F(t) + C \qquad \text{(A.25)}$$

In words, given a function $f(t)$, its integral is another function $F(t)$, plus a constant, where $dF(t) / dt = f(t)$. Integration in the form defined by equation (A.25) is called an *indefinite integral* because it does not show the limits over which the integration is to take place. When these limits are shown, we call the integral a *definite integral* and write it as

$$\int_{t=t_1}^{t=t_2} f(t)dt = \left[F(t) + C \right]_{t=t_1}^{t=t_2} = F(t_2) - F(t_1)$$

$$\text{(A.26)}$$

The limits over which integration is to take place are defined by t_1 and t_2. The constant of integration drops out.

Sometimes the best way to view integration is with graphs. For example, equation (A.25) indicates that $F(t)$ is equal to $\int f(t)dt$ minus the integration constant. For simplicity, assume the integration constant is zero and look at the following graphs:

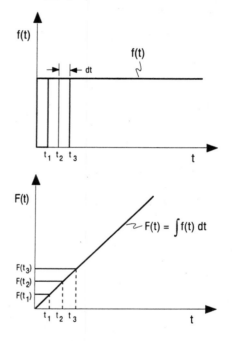

You can see that $F(t)$ is equal to the area under the $f(t)$ curve. Clearly, when t is very small, this area is zero. Each time t increases by an amount dt, the area under the curve increases a constant amount. Since $F(t)$ represents the area under the curve described by $f(t)$ from time $t = 0$ to time t, then $F(t_2) - F(t_1)$ must equal the area under the $f(t)$ curve from $t = 0$ to t_2, minus the area under the $f(t)$ curve from $t = 0$ to t_1. In graphical form:

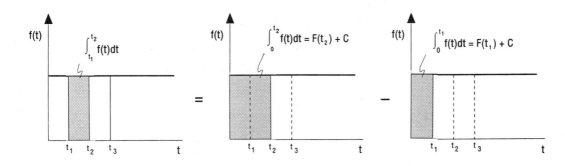

Generally, the independent variable is understood when writing the definite integral and only the limits are shown. That is,

$$\int_{t=t_1}^{t=t_2} f(t)dt = \int_{t_1}^{t_2} f(t)dt = \left[F(t)+C\right]_{t=t_1}^{t=t_2}$$
$$= \left[F(t)+C\right]_{t_1}^{t_2} = F(t_2) - F(t_1)$$

are all equivalent. *Notice that a definite integral is a function of its limits, not a function of the dependent variable t.*

As with differential calculus, integration formulas are available for finding the integral of many functions. I found that I saved a lot of time by committing to memory the formulas given in Table A.3.

You can also double-integrate a function just as you can double-differentiate a function. That is,

$$\iint f(t)dt = F(t) + C_1 t + C_2 \qquad (A.27)$$

Two constants of integration must now be evaluated. For example, in the car example we can take $f(t)$ as

$$f(t) = a = 10\,ft/\sec^2$$

That is, $f(t)$ is a constant. Double-integrating to get x gives

$$x = \iint 10 dt = 5t^2 + C_1 t + C_2 \qquad (A.28)$$

We know that at $t = 0$, $x = 0$ and $v = 0$. Therefore

$$x = 5\cdot 0^2 + C_1 \cdot 0 + C_2 = 0 \qquad \Rightarrow C_2 = 0$$
$$(A.29)$$

and

$$v = 10t_1 + C_1 = 10\cdot 0 + C_1 = 0 \Rightarrow C_1 = 0$$
$$(A.30)$$

We arrive at

$$x = 5t^2$$

as before.

I introduced you earlier to the differentiation operator D = d/dt. The inverse *integration operator* is

$$\frac{1}{D} = \int f(t)dt$$

**Table A.3. Most frequently used integrals
(where *c* and *a* are constants and
u and *v* are functions of *x*).**

$$\int du = u + c \qquad \int u^{-1}du = \int \frac{du}{u} = \ln|u| + c \qquad \int \sec u \tan u \, du = \sec u + c$$

$$\int (du + dv) = \int du + \int dv \qquad \int \sin u \, du = -\cos u + c \qquad \int \csc u \cot u \, du = -\csc u + c$$
$$= u + v + c$$

$$\int \cos u \, du = \sin u + c$$

$$\int adu = a\int du = au + c \qquad\qquad\qquad\qquad\qquad \int a^u du = \frac{a^u}{\ln a} + c$$

$$\int \sec^2 u \, du = \tan u + c$$

$$\int u^n du = \frac{u^{n+1}}{n+1} + c \qquad\qquad\qquad\qquad\qquad \int e^u du = e^u + c$$
$$\text{if } n \neq -1 \qquad \int \csc^2 u \, du = -\cot u + c$$

and in block form

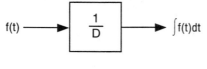

Example:

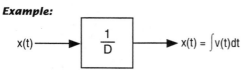

An "integrator" BASIC computer program is given in Listing A.3 and a spreadsheet version is given in Listing A.4. Carefully review these programs, and experiment with both of them. Change the number of steps required to compute the integral. Notice that you get a different answer each time. Remember that both of these programs give approximations to the integral and are based on equation (A.18). As Δt gets smaller, the answer you get will better approximate the correct value.

Listing A.3. BASIC integrator.

```
5     REM BASIC INTEGRATOR

10    DEF FN X(T) = 10*T^2

20    INPUT "VALUE OF UPPER LIMIT, T2,
      PLEASE";T2

30    INPUT "VALUE OF LOWER LIMIT,
      T1, PLEASE";T1

40    INPUT "NUMBER OF INTEGRATION
      STEPS, PLEASE";N

50    DELT = (T2-T1)/N

60    SUM = 0: T = T1

70    FOR I = 1 TO N

80    SUM = SUM + DELT*FN X(T)

90    T = T + DELT

100   NEXT

110   PRINT "VALUE OF INTEGRAL
      IS";SUM

120   END
```

A.4 Partial Derivatives

The equation for the volume V of a cylinder shows that it is a function of two variables, r and h. That is,

$$V = V(r,h) = \pi r^2 h$$

As I indicated earlier, you can hold one independent variable constant in this equation and then investigate the effect the other has on the dependent variable. You can do this regardless of how many independent variables there are in an equation. In essence, you convert a multiple independent variable function into a single variable function. You can then take derivatives of this function just as you would for a function that had only one independent variable. Such derivatives are called *partial derivatives* and the symbol $\partial u / \partial x$ is used to denote the partial derivative of u with respect to x.

For the general function $u = f(x,y)$, the first partial derivatives are defined as

Listing A.4. Spreadsheet integrator.

	A	B	C	D
1	T1			
2	T2			
3	N			
4	DELT	=(B2-B1)/B3		
5				
6	T	X(T)	X(T)*DELT	SUM
7	=B1	=10*A(7)^2	=B7*B4	=0
8	=A7+B4	=10*A(8)^2	=B8*B4	=D7+C8

$$\frac{\partial u}{\partial x} = \lim_{\Delta x \to 0} \frac{f(x + \Delta x, y) - f(x, y)}{\Delta x} \quad \text{(A.31)}$$

and

$$\frac{\partial u}{\partial y} = \lim_{\Delta y \to 0} \frac{f(x, y + \Delta y) - f(x, y)}{\Delta y} \quad \text{(A.32)}$$

You can see that these definitions are essentially identical to the definition previously given for a function of a single variable. Because the function has two independent variables, there are now two partial derivatives.

Let's take the partial derivatives of the equation for the volume of a cylinder. First, the partial derivative of V with respect to r is:

$$\frac{\partial V}{\partial r} = \lim_{\Delta r \to 0} \frac{V(r + \Delta r, h) - V(r, h)}{\Delta r}$$

$$= \lim_{\Delta r \to 0} \frac{\pi(r + \Delta r)^2 h - \pi r^2 h}{\Delta r}$$

$$= \lim_{\Delta r \to 0} \frac{\pi r^2 h + 2\pi r \Delta r h + \pi(\Delta r)^2 h - \pi r^2 h}{\Delta r}$$

$$= \lim_{\Delta r \to 0} 2\pi r h + \pi \Delta r h$$

$$= 2\pi r h \quad \text{(A.33)}$$

Next take the partial derivative of V with respect to h:

$$\frac{\partial V}{\partial h} = \lim_{\Delta h \to 0} \frac{V(r, h + \Delta h) - V(r, h)}{\Delta h}$$

$$= \lim_{\Delta h \to 0} \frac{\pi r^2 (h + \Delta h) - \pi r^2 h}{\Delta h}$$

$$= \lim_{\Delta h \to 0} \frac{\pi r^2 h + \pi r^2 \Delta h - \pi r^2 h}{\Delta h}$$

$$= \lim_{\Delta h \to 0} \frac{\pi r^2 \Delta h}{\Delta h}$$

$$= \pi r^2 \quad \text{(A.34)}$$

You can see that taking a partial derivative simply involves treating one variable as if it were a constant. You don't have to use the definition equations to compute the partial derivatives; simply recall or refer to the differentiation formulas for a single variable function given in Table A.2.

You can also take higher partial derivatives of multivariable functions. The partial derivatives are written as:

$$\frac{\partial}{\partial x}\left(\frac{\partial u}{\partial x}\right) = \frac{\partial^2 u}{\partial x^2} \quad \text{(A.35)}$$

$$\frac{\partial}{\partial y}\left(\frac{\partial u}{\partial x}\right) = \frac{\partial^2 u}{\partial x \partial y} = \frac{\partial}{\partial x}\left(\frac{\partial u}{\partial y}\right) \quad \text{(A.36)}$$

$$\frac{\partial}{\partial y}\left(\frac{\partial u}{\partial y}\right) = \frac{\partial^2 u}{\partial y^2} \quad \text{(A.37)}$$

Let's take these higher partial derivatives for the volume of a cylinder:

$$\frac{\partial}{\partial r}\left(\frac{\partial V}{\partial r}\right) = \frac{\partial^2 V}{\partial r^2} = 2\pi h \quad \text{(A.38)}$$

$$\frac{\partial}{\partial h}\left(\frac{\partial V}{\partial r}\right) = \frac{\partial^2 V}{\partial h \partial r} = \frac{\partial}{\partial r}\left(\frac{\partial V}{\partial h}\right) = 2\pi r \quad \text{(A.39)}$$

$$\frac{\partial}{\partial h}\left(\frac{\partial V}{\partial h}\right) = \frac{\partial^2 V}{\partial h^2} = 0 \quad \text{(A.40)}$$

Increments, Differentials and Total Derivatives

We previously defined the incremental change Δy in a function $y = f(x)$ of a single independent variable x as

$$\Delta y = f(x + \Delta x) - f(x)$$

When Δx is small, the increment Δy is essentially the same as the differential df. So for small values of Δx, we can write

$$\Delta y \cong dy = \frac{df}{dx} dx = \frac{df}{dx} \Delta x \qquad (A.41)$$

For a function $u(x,y)$ of two independent variables the increment Δu is

$$\Delta u = f(x + \Delta x, y + \Delta y) - f(x, y)$$

$$= \left[f(x + \Delta x, y + \Delta y) - f(x, y + \Delta y) \right]$$
$$+ \left[f(x, y + \Delta y) - f(x, y) \right]$$

$$= \frac{\partial u}{\partial x} \Delta x + \frac{\partial u}{\partial y} \Delta y \qquad (A.42)$$

As before, we can replace Δx with dx and Δy with dy and write the total differential as

$$du = \frac{\partial u}{\partial x} dx + \frac{\partial u}{\partial y} dy \qquad (A.43)$$

Let's now apply these equations to the volume of a cylinder. We can write

$$\Delta V = \frac{\partial V}{\partial r} \Delta r + \frac{\partial V}{\partial h} \Delta h \qquad (A.44)$$

Substituting the partial derivatives from above gives

$$\Delta V = (2\pi rh)\Delta r + \left(\pi r^2\right)\Delta h \qquad (A.45)$$

The application of this latter equation should now be clear. If "operating point" values for r and h, say r_o and h_o, are chosen, then this last equation provides the incremental change in volume of the cylinder as a function of the incremental changes in the radius and the height about the operating point r_o, h_o That is,

$$\Delta V = \left(2\pi r_o h_o\right)\Delta r + \left(\pi r_o^2\right)\Delta h \qquad (A.46)$$

You will note that this latter equation is linear in Δr and Δh.

A.5 Taylor's Theorem

You may recall from your algebra that any continuous function $y = f(x)$ can be expanded into an infinite series. We can restrict our attention to a point, $x = a$, and expand the function $f(x)$ about this point in the form

$$f(x) = b_o + b_1(x - a) + b_2(x - a)^2$$
$$+ \ldots + b_n(x - a)^n \qquad (A.47)$$

The coefficients for this equation can be found by taking successive derivatives and then evaluating the derivatives at $x = a$. That is,

$$\frac{df(x)}{dx} = b_1 + 2b_2(x-a) + \ldots + nb_n(x-a)^{n-1} \tag{A.48}$$

$$\frac{d^2f(x)}{dx^2} = 2b_2 + \ldots + n(n-1)b_n(x-a)^{n-2} \tag{A.49}$$

and so forth. Evaluating these at $x = a$ gives

$$f(a) = b_o \tag{A.50}$$

$$\left.\frac{df(x)}{dx}\right|_{x=a} = b_1 \tag{A.51}$$

$$\left.\frac{d^2f(x)}{dx^2}\right|_{x=a} = 2b_2 \tag{A.52}$$

and so forth. Substituting these values back into the expression for $f(x)$ gives

$$f(x) = f(a) + \left.\frac{df(x)}{dx}\right|_{x=a} \times (x-a) + \frac{1}{2}\left.\frac{d^2f(x)}{dx^2}\right|_{x=a}$$
$$\times (x-a)^2 + \ldots + \frac{1}{n!}\left.\frac{d^nf(x)}{dx^n}\right|_{x=a} \times (x-a)^n \tag{A.53}$$

This equation is known as Taylor's Series and it can be proven that the series converges.

One of the most important applications of this equation is associated with the linearization of functions. If we let $\Delta x = x - a$, and use only the first two terms of the Taylor's Series, then any function can be approximated by

$$f(x) = f(a) + \left.\frac{df(x)}{dx}\right|_{x=a} \times \Delta x \tag{A.54}$$

Taylor's Series can also be used with multi-variable functions. The function $u(x,y)$ can expand about a point (x_o, y_o). Then a linear approximation of the function would be

$$u(x,y) = u(x_o, y_o) + \left.\frac{\partial u(x,y)}{\partial x}\right|_{\substack{x=x_o \\ y=y_o}}$$
$$\times \Delta x + \left.\frac{\partial u(x,y)}{\partial x}\right|_{\substack{x=x_o \\ y=y_o}} \times \Delta y \tag{A.55}$$

You can see that this is equivalent to the total derivative given by equation (A.44) earlier.

APPENDIX B

DSP Vendors

Mathematical Tool Vendors

MathSoft Inc (makers of MathCAD)
    101 Main Street
    Cambridge, MA 02142
    (800) 628-4223
    e-mail: sales-info@mathsoft.com
    Web address: www.mathsoft.com

MathWorks, Inc. (makers of MatLab)
    24 Prime Park Way
    Natick, MA 01760
    (508) 653-1415
    e-mail: info@mathworks.com
    Web address: www.mathworks.com

Wolfram Research, Inc. (makers of Mathematica)
    100 Trade Center Drive
    Champaign, IL 61820
    (800) 441 -6284
    e-mail: info@wolfram.com
    Web address: www.wolfram.com

DSP Chip Vendors

Ariel Corporation (headquarters)
    2540 Route 130
    Cranbury, NJ 08512
    (609) 860-2900
    fax: (609) 860-1155
    e-mail: ariel@ariel.com

Analog Devices
    1 Technology Way
    Norwood, MA 02062
    (617) 461-3881
    Web address: www.analog.com

AT&T Microelectronics
    555 Union Blvd.
    Dept. AL500404200
    Allentown, PA 18103
    (800) 372-2447
    Web address: www.att.com

Motorola DSP Division
    6501 William Cannon Dr. W.
    Austin, TX 78735
    (512) 891-2030

Motorola Semiconductor Products Sector
    Communications & Advanced Consumer Technology
       Group
    Austin, Texas
    e-mail: dsphelp@dsp.sps.mot.com
    Web address: www.mot.com/SPS/DSP (great DSP site!)

NEC Electronics
>   475 Ellis Street
>   Mountain View, CA 94039
>   (415) 965-6159

Pentek, Inc.
>   55 Walnut Street
>   Norwood, NJ 07648
>   (201) 767-7100
>   fax: (201) 767-3994
>   e-mail: rodger@pentek.com

Texas Instruments, Semiconductor Group
>   P.O. Box 1712228
>   Denver, CO 80217
>   (800) 477-8924
>   Web address: www.ti.com
>   Check out TI's on-line DSPLab at www.dsplab.com
>   Customer Response Center: (800) 336-5236

White Mountain DSP, Inc.
>   Suite 433
>   131 DW Highway
>   Nashua, NH 03060-5245
>   (603) 883-2430
>   fax: (603) 882-2655
>   e-mail: info@wmdsp.com

Board-level Products

Communication Automation & Control, Inc
    1642 Union Blvd.
    Suite 200
    Allentown, PA 18103

CSPI (VME boards)
    40 Linnel Circle
    Billerica, MA 01821
    (800) 325-3110

Data Translation (PC, PCI)
    100 Locke Drive
    Marlboro, MA 01752
    (508) 481-3700
    e-mail: Info@datx.com
    Web address: www.datx.com

DSP Research, Inc. (PC, PCI)
    1095 East Duane Avenue
    Suite 203
    Sunnyvale, CA 94086
    (408) 773-1042

National Instruments (PC)
    6504 Bridge Point Parkway
    Austin, TX 78730
    (800) 443-3488
    e-mail: info@natinst.com
    Web address: www.natinst.com

SONITEC International Inc. (PC)
    14 Mica Lane
    Wellesly, MA 02181
    (617) 235-6824

White Mountain DSP (PC)
    131 DW Highway
    Suite 433
    Nashua, NH 03060-5245
    (603) 883-2430

APPENDIX C

Useful Magazines and Other Publications

Communication Automation & Control, Inc.
    1642 Union Boulevard, Suite 200
    Allentown, PA 18103-1585
    (610) 776-6669
    e-mail: sales@cacdsp.com
    Web address: www.cacdsp.com

Communication Systems Design
    Monthly, devoted to communications, multimedia,
        and DSP.
    FREE to qualified engineers.
    (415) 905-2200
    Web address: www.csdmag.com

Communications Week
    CMP publication with lots of graphics.
    (516) 562-5000
    techweb.cmp.com/cw/current

Computer Design
    Monthly, articles on computer design, DSP, embedded
        systems, etc.
    FREE to qualified engineers.
    (603) 891-0123
    Web address: www.computer-design.com

DSP and Multimedia Technology
    Bi-monthly, paid circulation.
    (415) 969-6920
    na.htm

IEEE Signal Processing Magazine
> IEEE's flagship magazine for DSP and signal processing issues.
> Available to IEEE members and non-members.
> (212) 705-7900

Personal Engineering and Instrumentation News
> Mark Sullivan writes a DSP column for this magazine.
> He provides downloadable source code on the Web at www.access.digex.net/~dalek.

Tech Central
> Online product mart devoted to DSP, embedded systems, machine vision, and other areas. Registration required.
> Web address: www.techcentral.com

Tech Online
> Another online product mart.
> Web address: www.techonline.com

Additional Web Resources

Amateur Radio DSP Page
> Web address: www.tapr.org/dsp/index.html

DSP Internet Resource List
> Web address: www.cera2.com/dsp.htm

Glossary

Analog-to-Digital Converter (ADC) — Converts an analog voltage into a digital number. There are a number of different types, but the most common ones found in DSP are the Successive Approximation Register (SAR) and the Flash converter.

Analog Frequency — The analog frequency is what we normally think of as the frequency of the signal. *See Digital Frequency.*

Anti-Aliasing Filter — A filter that is used to limit the bandwidth of any incoming signal.

Digital Signal Processing (DSP) — As the term states, this is the use of digital techniques to process signals. Examples include the use of computers to filter signals, enhance music recordings, study medical and scientific phenomena, create and analyze music, and numerous other related applications.

Digital-to-Analog Converter (DAC) — Converts a digital number to an analog voltage.

Digital Frequency — The digital frequency is the analog frequency scaled by the sample interval. If λ is the digital frequency, f is the analog frequency, and T is the sample period, then $\lambda = f/T$. The digital frequency is normally expressed over the range of $-\pi$ to π. See *Analog Frequency.*

Discrete Fourier Transform (DFT) — A computational technique for computing the transform of a signal. Normally used to compute the spectrum of a signal from the time domain version of the signal. See *Inverse Discrete Fourier Transform (IDFT)*, *Fourier Transform*, and *Fast Fourier Transform (FFT)*.

DSP Processor — DSP processors are specialized to perform computations in a very fast manner. Typically, they have special architectures that make moving and manipulating data more efficient. Typically, DSP processors have both hardware and software features that are optimized to perform the more common DSP functions (convolution, for example.)

Fast Fourier Transform (FFT) — Computationally efficient version of the *Discrete Fourier Transform*. The FFT is based on eliminating redundant computations often found in processing the DFT. For large transforms, the FFT may be thousands of times faster than the equivalent DFT. See *Inverse Discrete Fourier Transform (IDFT)*, *Fourier Transform*, and *Fast Fourier Transform (FFT)*.

Finite Impulse Response Filters (FIR) — A filter whose architecture guarantees that its output will eventually return to zero if the filter is excited with an impulse imput. FIR filters are unconditionally stable. See *Infinite Impulse Response Filter*.

Fourier Transform — A mathematical transform using sinusoids as the basis function. See the *Discrete Fourier Transform (DFT)* and the *Fast Fourier Transform (FFT)*.

Fourier Series — A series of sinusoid wave forms that, when added together, produce a resultant wave form.

Harvard Architecture — A common architecture for DSP processors, the Harvard architecture splits the data path and the instruction path into two separate streams. This increases the parallelism of the processor, and therefore improves the throughput. See *DSP Processors*.

Infinite Impulse Response Filters (IIR) — A filter that, once excited, may have an output for an infinite period of time. Depending upon a number of factors, an IIR may be unconditionally stable, conditionally stable, or unstable.

Inverse Discrete Fourier Transform (IDFT) — A computational technique for computing the transform of a signal. Normally used to compute the time domain representation of a signal from the spectrum of the signal. See *Discrete Fourier Transform (DFT)*, *Fourier Transform*, and *Fast Fourier Transform (FFT)*.

Smoothing filter — A filter that is used on the output of the DAC in a DSP system. Its purpose is to smooth out the stair step pattern of the DAC's output.

Von Neumann Architecture — The standard computer architecture. A Von Neumann machine combines both data and instructions into the same processing stream. Named after mathematician Johaan Von Neumann (1903–1957), who conceived the idea.

Window — As applied to DSP, a window is a special function that shapes the transfer function. Typically used to tweak the coefficients of filters.

References

1. Foster, Caxton C., *Real Time Programming*, ISBN 0-201-01937-X, Addison Wesley Publishing Company, Inc., 1981

2. Peled, Abraham , and Liu, Bede, *Digital Signal Processing*, ISBN 0-471-01941-0, John Wiley and Sons, Inc., 1976

3. Rorabaugh, C. Britton, *Digital Filter Designer's Handbook*, ISBN 0-07-911166-1, McGraw-Hill, 1993

4. Smith, Mark J.T., and Mersereau, Russell M., *Introduction to Digital Signal Processing*, ISBN 0-471-51693-7, John Wiley and Sons, Inc., 1992

5. Stanley, Willam D., *Network Analysis with Applications*, ISBN 0-8359-4880-3, Reston Publishing Company, Inc., 1985

6. Stearns, Samuel D., *Digital Signal Analysis*, ISBN 0-8104-5828-4, Hayden Book Company, 1975

7. Willams, Charles S., *Designing Digital Filters*, ISBN 0-13-20186-X, Prentice Hall, 1986

Index